SUPER SPECIES

the creatures that will dominate the planet

SUPER SPECIES

the creatures that will dominate the planet

Garry Hamilton

FIREFLY BOOKS

A FIREFLY BOOK

Published by Firefly Books Ltd. 2010

First printing

Publisher Cataloging-in-Publication Data (U.S.)

Hamilton, Garry.
    Super species : the creatures that will dominate the planet / Garry Hamilton.

[272] p. : col. photos. ; cm.
Includes bibliographical references and index.
Summary: Profiles invasive species that are dominating ecosystems around the world, including the adaptive traits of invasive species, the methods through which species spread to new territories, environmental damage and scientific viewpoints.
ISBN-13: 978-1-55407-630-7
ISBN-10: 1-55407-630-7
1. Biological invasions. 2. Evolution (Biology). I. Title.
577/.18 dc22   QH353.H365 2010

Library and Archives Canada Cataloguing in Publication

Hamilton, Garry
    Super species : the creatures that will dominate the planet / Garry Hamilton.

Includes bibliographical references and index.
ISBN-13: 978-1-55407-630-7
ISBN-10: 1-55407-630-7
1. Introduced organisms. 2. Biological invasions. I. Title.
QH353.H34 2010   578.6'2   C2010-901827-3

Published in the United States by
Firefly Books (U.S.) Inc.
P.O. Box 1338, Ellicott Station
Buffalo, New York 14205

Published in Canada by
Firefly Books Ltd.
66 Leek Crescent
Richmond Hill, Ontario L4B 1H1

Cover and interior design: Kimberley Young

Printed in China

The publisher gratefully acknowledges the financial support for our publishing program by the Government of Canada through the Canada Book Fund as administered by the Department of Canadian Heritage.

For Cecilia

CONTENTS

 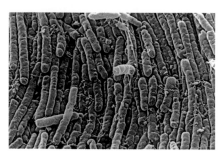

INTRODUCTION

I N RECENT YEARS CANADIAN fishermen off the coast of British Columbia have been encountering a creature that, not long ago, none of them had ever seen before — a species of large squid with bizarre behaviors, including a tendency to hunt in large, feverish packs that get so riled up individuals sometimes end up eating themselves. In Europe a type of ant formerly known only in South America has taken over much of the continent, partly because of its ability to form massive supercolonies that span thousands of square miles and overwhelm any rivals in their path. In the world's oceans a tiny jellyfish-like creature has spread around the globe, and scientists have recently uncovered evidence that it may be able to reverse the process of aging, extending its lifespan potentially forever.

Strange happenings indeed, and that's just the beginning. Rapidly expanding populations of feral pigs, taking advantage of the high reproductive rate bred into them on the farm, are reshaping habitats from Australia to Texas. Eastern gray squirrels, which thrive in environments disturbed by human activity, have swarmed over the British Isles and are currently advancing across Europe. A mysterious fungus has spread across most of the planet, decimating amphibian populations everywhere it goes. Scientists have even discovered microorganisms running amok in the ecosystem of tiny bugs found in your intestines. The list goes on and on. Everywhere you turn, it seems life is out of control — in the oceans, in freshwater lakes and rivers, in wetlands, in forests, in our own backyards and even in our guts.

You may recognize these upstarts as invasive species, a term commonly used to describe plants, animals and microbes that invade territories outside their native range. Sometimes they have merely expanded their historical range, but more often these "nonnative" organisms have been introduced by humans, either by accident or on purpose, into a new environment. Crops, garden plants and farm animals, for example, are often alien species. If they escape and establish self-reproducing populations, they join the ranks of the invaders.

The spread of invasive species around the world that has accompanied the rise of *Homo sapiens* — and continues to occur today at an accelerating pace — represents a reshuffling of the biosphere that is unprecedented in the history of life on Earth, or at least in the records that have been left behind as fossils. In many habitats around the world, researchers have recorded dozens and even hundreds of thriving alien species. Some of these invaders have become the dominant species in their environment, and in the most dramatic instances, their presence has altered physical habitats and restructured local ecosystems.

One reason for what's happening is the fact that species now have more opportunities to spread than ever before. Mountains, rivers, oceans and continents once stood as insurmountable obstacles, so communities of organisms evolved in isolation from one another. Geological events may have broken down some of the barriers, but such occurrences are few and far between. For thousands of years humans have carried species with them in their travels, often to provide food and other resources. But the transported species often include unseen hitchhikers — diseases, pests, weeds and species that just happened to come along for the ride. As human populations have grown and our trade networks have expanded, the rate at which new species are being introduced into new places has accelerated.

Another driving force behind the unprecedented rise of invasive species is our changing environment. Invasive species are ecological opportunists. They are quick to take advantage of environmental disturbances, and if there's one thing humans are good at, it's disturbing environments. We've dammed rivers, polluted lakes and cut down forests. We've drained wetlands. We've altered how often fires burn across landscapes. We've removed almost all the top predators that once dominated continental and marine food chains. We've covered terrestrial habitats with roads and developed tree farms, orchards and fields planted with acre after acre of single-species crops. We've changed the vegetation of the Great Plains of North America through livestock grazing. We've turned lakes and rivers into toxic soups with the chemicals that flush from our farms into watersheds with every downpour of rain. We've built cities and towns. We've altered the chemical composition of air, soil and water. We're changing the pH level of the oceans. We've tinkered with the amount of ultraviolet radiation that reaches Earth's surface. And with the buildup of greenhouse gases in the upper atmosphere, we're possibly altering global temperatures.

In many of these cases we've not only disturbed existing habitats, we've converted them into entirely new ones that nature has never seen before, everything from concrete-sided canals to inner-city parking lots to effluent ponds outside refineries. As our population has increased, so too has our impact on the environment. With each step we've created new ecological opportunities, and invasive species have followed close behind, taking advantage of them.

Finally, their spread is due in large part to the characteristics of the invasive species themselves. Of all the life-forms that have the potential to travel around the world, and of all those that are known to have done so, only a small fraction have been able to survive in their new locations. And only a small fraction of those have been able to thrive as seemingly out-of-control invaders. Almost invariably they come from comparatively humble origins — as previously unheralded life-forms struggling for survival on the floodplains of Argentina, in the brackish seas of Central Asia or the rivers of New Zealand. These species have been able to adapt and win the struggle for survival in a world turned upside down. Often they seem like freaks of nature. They can be unusually aggressive, hard to kill, unfazed by the activities of humans, capable of astonishingly rapid growth and reproduction, tolerant of pollution — or all of the above. But mainly they're just phenomenally successful. Call them Super Species: the increasingly dominant life-forms that are poised to inherit Earth.

Most ecologists and conservationists believe that invasive species now represent one of our biggest problems, high up on a list that includes issues such as global warming. Out-of-control invasive species are a source of major headaches for humans. Rampant plants, animals, insects and microorganisms are causing enormous expense as they ruin crops, interfere with business operations and spread disease. The costs associated with these problems, which are borne first by those directly in the path of the invading species and then spread to the rest of us, are so great that many now regard them as a drain on the global economy and a threat to current standards of living. In 2005 an American study tallied the known costs associated with invasive species: in the United States alone, the amount came to $120 billion a year.

And that's just economic cost. Many scientists believe that invasive species represent a major threat to the global environment. They regard the proliferation of nonnative species as a direct threat to native varieties. They see them as the driving force behind the potentially irreversible collapse of long-standing ecosystems in environments around the world: in aquatic habitats such as the Mediterranean Sea, the Great Lakes and Africa's Lake Victoria, and in terrestrial habitats such as the rain forests of Hawaii and the rangelands of western North America. Most ecologists now accept as a matter of fact that invasive species are the second-greatest threat to biodiversity next to habitat loss, a cancer that is eating away at Earth's biosphere.

But some ecologists are beginning to see the changes sweeping across the planet in a much different light. To them the rise of invasive species is not so much a disease as a symptom of our own impacts on the planet. According to this perspective, native ecosystems are collapsing not because invaders are driving away native species but because humans have altered or destroyed the conditions under which the ecosystems first evolved. Invasive species are simply opportunists taking advantage of the resulting chaos, filling the voids created by our presence. And some researchers have suggested that invasive species may even represent a cure: nature's attempt to rebuild ecological vibrancy in the wake of human-mediated destruction. This idea is now gaining traction, thanks to emerging evidence that invasive species can trigger positive effects as well as negative ones — both ecologically and as potential resources — and that their overall ecological and economic impacts aren't anywhere near as severe as predicted.

Right now the one certainty is that Earth's biosphere is changing. No one knows yet whether invasive species can help rebuild ecosystems that support high degrees of diversity and provide the ecological services — cleaning the air and water, building fertile soil, sustaining food chains — that we depend on for survival, or whether they will turn the world into a toxic soup. All we know is that increasingly our destiny is becoming linked with the species now inheriting the world. They're the current winners in nature's ruthless process of natural selection. If life is indeed survival of the fittest, then the fittest have arrived. It's time we got to know the winners, the Super Species in our midst. For better or for worse, they are the new face of the biosphere.

Part One

NATURE RESHUFFLED

IF THE CURRENT TRANSFORMATION of the Earth's biosphere is a result of multiple factors, the first key factor is dispersal: species are spreading around the planet on a massive scale, largely because they now have more opportunities to do so. In the past, this wasn't so easy. Mountain ranges, deserts, oceans and rivers act as physical barriers that restrict the movement of even the most successful species. But widespread dispersal has also been constrained by the nature of life itself.

The diversity of the Earth's habitats — arising from regional variations in light, temperature, soil chemistry, salinity, altitude, latitude, seasonal patterns and moisture — and the slow pace of geological time have been ideal incubators for driving not only diversification of life but also a dependency between species and their ecological niches. In northern Australia and Central America there are frogs so highly specialized that their ranges are limited to only a few square miles of rain forest. Many freshwater mussels in the United States have a native range spanning no more than a few rivers. The Delmarva fox squirrel's range in the mature hardwood forests of the eastern U.S. originally included nothing more than a peninsula now shared by Delaware and Maryland, as well as a few bits of the adjoining states. Thus diversity becomes a barrier in its own right.

At the same time, biological invasion is also an undeniable fact of life. In 1883 the volcanic island of Krakatoa, which lies in the Indonesian archipelago between the islands of Java and Sumatra, erupted in one of the greatest displays of physical force in recorded history. The blast, which could be heard as far away as Perth, Australia, is believed to have sterilized the fragments of the island that remained above sea level. However, within months scientists were coming across spiders that had apparently floated there on the wind. Not long afterward, the first grasses appeared, and wandering birds brought the seeds of other plants in their droppings. Today the islands are covered in rain forest, with birds and bats and more than 300 different plant species, including 24 types of fig trees. The process of ecosystem renewal is still underway, thanks entirely to the lottery of natural invasion.

Similarly there have been periods throughout the history of Earth when the slow unfolding of geological processes — the movement of continental plates, the advance and retreat of glaciers — has opened the door to large-scale invasions. One such event occurred some three million years ago, when the isthmus of Panama rose above sea level, creating the first land bridge between North and South America. Among the species that used this opportunity to advance northward were giant ground sloths and terror birds, a group of flightless carnivorous birds that included species standing 10 feet (3 m) tall. Those particular invaders eventually died out, but other northbound migrants from that time — armadillos, porcupines and opossums — continue to inhabit North America to this day. The mixing went the other way too, with South America seeing an influx of life-forms that included ancestral bears, large predatory cats, various members of the wild dog family, horses, deer and llamas, all of which are believed to have had a major impact on the southern continent's ecosystem.

One additional example that shouldn't be left out is the land bridge across the Bering Strait, which has formed a connection between North America and Asia every time sea levels fall in response to the formation of ice-age glaciers. Over different periods, mammoths, deer and bison came this way to populate North America, as did what may be the most invasive of all species — *Homo sapiens*.

Despite this history, many scientists now believe that life on Earth is currently undergoing an unprecedented degree of mixing. Thanks to the emergence of humans as a conduit, all the barriers that previously isolated species can now be negotiated. And instead of select species in specific regions taking advantage of unusual dispersal opportunities on geological time scales, we are now witnessing an era in which a wide variety of plants, animals and microbes are crisscrossing the globe on a regular basis. Some people have described this reshuffling as the "McDonaldization" of nature — a transformation in which the diversity of life is being replaced by a homogeneous assembly of invasive species, in the same way that American culture has invaded and overwhelmed traditional lifestyles in other parts of the world.

Although commonly regarded as a recent phenomenon, the current dispersal of species probably dates back to when humans themselves first became invaders. After their emergence in Africa, our ancestors are thought to have spread throughout the Middle East, Eurasia and Australia between 70,000 and 50,000 years ago. Further expansion resulted in *H. sapiens* arriving in the Americas beginning around 15,000 years ago, and eventually in far-flung places such as the islands of Polynesia and the harsh frontiers of the Arctic. At almost every step these early pioneers are thought to have brought with them new species. Layers of bones excavated from caves in Indonesia and New Guinea reveal that human immigrants likely introduced possums, wallabies, rats, dogs and cassowaries. The tree-dwelling cuscus was brought to islands off New Guinea more than 10,000 years ago. Around 4,000 years ago, Asian seafarers are believed to have brought the first dingoes to Australia.

After the seafaring Polynesians began exploring and settling the islands of the Pacific 2,000 years ago, many different species were spread both intentionally (for food, shelter and medicine) and by accident. The list of animals these early explorers brought to destinations such as Hawaii includes pigs, dogs, Pacific rats, red jungle fowl (ancestors of the domestic chicken), several species of geckos and skinks, fleas, lice and houseflies. Introduced plants included important crop species such as banana trees, sugarcane, sweet potatoes and bamboo, as well as many crop-associated weeds. European colonists also took species with them, again by accident and by design. The former include various well-known rodents such as black rats, Norway rats and house mice, as well as some surprises. Those earthworms that you dig up in your backyard or use as fishing bait? They're European earthworms, descendents of stowaways that likely arrived with the first pioneers.

As for intentional introductions, early European settlers displayed a remarkable desire to surround themselves with whatever familiar plants and animals could survive in their new environment. Honeybees were imported for honey, pigeons for racing, foxes, red deer and rabbits for sport hunting, and various trout species for fishing. Nothing compared, however, to the exploits of the gardeners. British colonists sweltering in the tropical heat of the Indian subcontinent retreated to cool, moist highland hill stations, where they were able to surround themselves with familiar plants from home: buttercups, violets, wild strawberries, raspberries, chicory, deadnettle, blindeyes, creeping thyme. By the end of the 1800s, many of these species were growing wild alongside imported weeds such as hairy spurge and sow thistle.

Similar introductions occurred throughout the English colonies during the 1800s, thanks to the rise of acclimatization societies, which were natural history organizations whose members often saw it as their duty to import familiar species missing from the local ecosystems. Eugene Schieffelin, chairman of the American Acclimatization Society in the 1870s, was an avid theatre fan who made it his mission to introduce into the U.S. every bird mentioned in every play by Shakespeare. His greatest success was with the species mentioned in act 1, scene 3, of *Henry IV, Part 1* — the European starling. Around this time Western gardeners developed a love for exotic Asian plants, particularly species from China and Japan that grow well in European and North American climates. Kudzu vines, Japanese barberry, Oriental bittersweet, burning bush, Japanese honeysuckle, Amur corktree, buckthorn, Japanese knotweed, tree-of-heaven, princess trees and Amur maples eventually became common nursery items throughout the West. All have since become problem weeds.

Invasion isn't just a by-product of human immigration. It's a fundamental component of our nature, a side effect of our seemingly limitless ingenuity, our instinct for pursuing short-term goals at the expense of long-term consequences, and our attempts to bend the rest of nature to suit our needs. Consider agriculture and aquaculture, manifestations of our successful desire to gain control over the supply of food. Usually cultivated species — those bred more for human use than for survival in the wild, and almost always imported — remain where they're planted. But this isn't always true, especially in the case of animals. The past century has seen frogs escaping from frog farms, oysters from oyster farms, pigs from pig farms, fish from fish farms. Many of these breakouts have resulted in the establishment of nonnative populations in foreign environments.

"Farming" has frequently amounted to nothing more than introducing nonnatives into the wild and then harvesting the results. Goats were introduced to islands such as the Galapagos as a source of wild meat for visiting whalers. Arctic foxes were brought to the islands of Alaska as a source of fur. The ancient Romans introduced carp into western Europe as a food source. Since then humans have been moving various species of carp around the world and today these fish are nonnative residents in more than 120 countries. In the U.S., invasive silver and bighead carp — two species commonly known as Asian carp — have become so prevalent that in some places they literally jump into passing motorboats, and bow-hunting for flying Asian carp is becoming a popular American pastime.

In addition to direct introduction of crops and livestock into new habitats, agriculture is to blame for the spread of the many species associated with domesticated life-forms: the parasites, microbes and various other freeloaders that inevitably travel along with imported farm products. Importation of oysters for cultivation in the Mediterranean is thought to be the main reason why more than 60 different species of nonnative seaweed now inhabit that sea. The water mold responsible for potato blight, which during the mid-19th century caused a famine in Ireland that killed an estimated one million people, is believed to have been transported to the British Isles on potatoes brought from America.

Trade is another part of human activity that is responsible for widespread reshuffling of species. With the expansion of international trade and the global consumer society, more goods are being moved between more destinations at increasing speeds and in more ways, steadily increasing the list of organisms whose eggs or seeds or spores can potentially survive long journeys between continents. One way this happens is when species hitch rides along with cargo, particularly trees and other natural resources. Asian longhorned beetles, balsam woolly adelgids and gypsy moths are just a few of the pests that have spread widely via shipments of raw timber or hidden in wooden packing materials or in imported nursery plants and soil.

Another cause of spread is the common practice of ballast dumping. Ships sailing empty to collect trade goods from foreign ports often require some form of deadweight — soil, or more commonly today, water — for stability. Once they've arrived in port, the unwanted ballast — along with any seeds, clippings, spores, eggs, insects or other creatures it may contain — is discharged into the environment. Discarded ballast is believed to have spread a wide range of organisms, from purple loosestrife to zebra mussels to jellyfish to gobies, small central Asian fish that have invaded the Great

Lakes and parts of Europe. Soil ballast likely brought species such as the fire ant to America, where it has become a major pest across the southern and southwestern parts of the country.

Trade also leads to dismantling of the physical barriers that prevented natural species migrations for thousands of years. The construction of a system of locks and canals on the upper St. Lawrence River gave species such as the sea lamprey access to the Great Lakes. The Suez Canal, which opened in 1869, allowed mixing between marine species of the Red Sea (and, indirectly, the Indian Ocean) and the Mediterranean. The Red Sea residents have been particularly invasive: more than 300 species have now taken up residence in the Mediterranean. When it was completed in 1914, the Panama Canal created an even more complex mixing pot, bringing together two previously unconnected river systems while at the same time opening a link between the Pacific Ocean and the Caribbean Sea. Among the results: five species of nonnative fish have colonized the Rio Grande River on one side of the canal, while three other invasive fish species have made their way into the Chagres River system on the other side.

Species introductions have occurred because we humans often turn to nature to solve our problems. Settlers and ranchers replaced native vegetation with nonnative plants deemed better suited for grazing livestock. In Australia this included the prickly pear cactus, which became a major weed in the eastern part of the country after it was introduced from America in the early 1800s. Land managers in countries all over the world have repeatedly turned to hardy, fast-growing nonnative trees, shrubs and vines to control runaway soil erosion on hillsides clear-cut for timber harvesting. Many of the species used for this purpose — multiflora rose, salt cedar (also known as tamarisk), kudzu, crown vetch, Russian olive and black locust, to name just a few — are now troublesome invaders. Australian paperbark trees were brought to Florida to dry up swampland. Indian mongooses, cane toads and rosy wolfsnails are on the list of voracious predators that became invasive species after being imported to control other invasive species. Mosquitofish were spread throughout the tropics to control malaria. African clawed frogs were used for human pregnancy tests, then released into the wild when more sophisticated methods were developed.

Many introductions occur because of human carelessness or thought-lessness (or both). Discarded fishing bait has resulted in the spread of crayfish. Cleaning out fish tanks has led to the introduction of nonnative seaweed and possibly the lionfish, a colorful predator with venomous spines that is currently terrorizing reef fish in the Caribbean. Burmese pythons — constrictors that can grow up to 19 feet (6 m) long and weigh up to 200 pounds (90 kg) — are now breeding in the Florida Everglades, thanks to fickle pet owners. The gypsy moth, reviled throughout eastern North America as a major defoliator of hardwood trees, was brought to the Boston area by a French astronomer who hoped to establish a commercial silk industry. Careless farming operations, poorly run zoos and inattentive pet owners have been responsible for the spread of oysters, bullfrogs, pigs and jungle mynas. Tourists, out of either ignorance or indifference, contribute to the process when they return from vacation with exotic foods or even fancy shells containing live mollusks.

While the term "invasive species" often suggests farm weeds or voracious predators, it's important to remember that the phenomenon also includes microbes. When Europeans arrived in the New World, they brought with them an invasive species that was among the smallest life-forms on the planet: the virus that causes smallpox. Other notorious invasive microbes include the fungi responsible for Dutch elm disease, which caused a 20th-century epidemic that decimated elm forests in Europe and North America, and chestnut blight, which is thought to have killed four billion American chestnut trees between 1900 and 1940. They also include the virus rinderpest, which decimated various African ungulates after it arrived with the first domestic cattle, and water molds such as potato blight. Often an invasive host is the cause of the invasive disease. Introduced cattle brought the rinderpest virus to Africa; invasive rats carried invasive fleas that carried *Yersinia pestis*, the bacterial source of bubonic plagues; Asian tiger mosquitoes, hitchhiking as eggs in water that had collected inside internationally traded used tires, brought dog heartworm disease to the United States.

In some ways, times have changed. Intentional introduction of wild plants and animals to new locations, particularly mammals and birds, is long out of fashion. The use of animals and microbes as biological control agents is in most places now carefully regulated to avoid unintentional escapes. A general rise in cultural preferences for natives over nonnatives has led to stricter laws governing which species can and cannot be trafficked across borders, and in some parts of the world customs inspectors now place a high priority on blocking the entry of unwanted species. But species continue to spread for a couple of reasons. For one thing, the number of introduced species that have successfully formed new populations is astonishingly large. By one estimate, at least 6,600 nonnative species have become established in North America since the arrival of Europeans. More than 250 nonnatives are thought to inhabit San Francisco Bay alone, and nearly half the plant species now found in New Zealand are exotics. Most parts of the world have been completely overrun by thriving invaders.

What's more, there's no indication that humans can control the accidental spread of exotic species. A study from Europe in the late 1990s showed that, despite careful controls designed to limit the introduction of invasive organisms, one new species was being introduced onto the continent *every three weeks*. On top of this, some nations seem less eager to adopt stringent control measures than others. Conservationists hailed the International Convention for the Control and Management of Ships' Ballast Water and Sediments as a key breakthrough when it was first drawn up in 2004. However, ratification requires backing by 30 nations, and by the end of 2009 only 19 countries were on board as signatories. And so the Great Reshuffling, as it's now known in some circles, continues.

AMERICAN BULLFROG

Lithobates catesbeianus

C ANNIBALISM HAS LONG BEEN a subject of mystery to biologists. In the early days some prominent theorists argued that it is a rare and aberrant behavior, something that happens when animals are cooped up in a lab cage, starving. That makes sense. Is there anything more unnatural, more contrary to the laws of survival, than the act of consuming one's own kind?

However, in recent decades the discovery that cannibalism occurs in many species under natural circumstances has led scientists to reconsider the question. Some now think that certain animals may rely on cannibalism, not only to survive hard times and possibly avoid extinction but also to regulate and stabilize population size, reduce the prevalence of disease and improve the overall fitness of both individuals and the population as a whole. It might even be a genetic trait. Although still speculative, this view has become increasingly harder to dispute, and one reason is the rise of the American bullfrog, *Lithobates catesbeianus* (formerly *Rana catesbeiana*). This highly invasive species is one of the world's most successful amphibians, partly because bullfrogs will eat just about anything they can swallow — including, to no small degree, their fellow bullfrogs.

When most people think "alien amphibians," what usually springs to

◀ A carnivorous predator, the bullfrog will eat just about any creature it can swallow. The ability to find, catch and digest a wide variety of prey is one of the main reasons why the species has become such a successful invader.

mind is the cane toad, a species that wreaked havoc after being introduced for pest control in places such as Fiji and northern Australia. But the bullfrog is enjoying even greater success. Originally native to the east coast of North America, including parts of Mexico and Canada, bullfrogs are now widely distributed throughout western North America, South America, Europe and Asia. What's more, recent reports suggest that the invasion is intensifying on several fronts. Humans are largely to blame, having shipped bullfrogs around the world for use as biological control agents, as pets or for sport (they're often the species of choice in frog-jumping competitions) and as additions to backyard ponds. But the primary driving force has been a practice that may strike most North Americans as surprising — frog farming.

Yes, despite the fact that many people cringe at the thought of eating swamp creatures, and even though amphibians aren't amenable to domestication, for centuries humans have made a living by raising frogs for human consumption. *L. catesbeianus* is the largest frog in North America, capable of growing to nearly 8 inches (20 cm) long and sometimes weighing more than a pound (0.5 kg). Its size is just one reason why budding capitalists might look at those big, meaty legs and see dollar signs. In the late 1800s, after gold miners in California had eaten their way through the native frogs, eager entrepreneurs imported American bullfrogs from back east to satisfy the increasing demand, and that was just the start. Bullfrogs were introduced into British Columbia for frog farming in the 1930s and '40s, and in locations all over the world right up to the 1990s. As of 2006, according to the United Nations Food and Agricultural Organization, commercial bullfrog operations were operating in at least 13 countries. Although statistics are rarely gathered, a recent study estimated that the number of bullfrogs imported into the U.S. between 1998 and 2002 approached nearly four million.

But profitable frog farming, as it turned out, isn't easy. This was particularly true during the early decades, when aquaculture techniques were still in the early stages of development. One problem is that bullfrogs are carnivorous predators, genetically programmed to eat only food that moves. This means that if you toss them cheap mass-produced food pellets — the secret to success in many large-scale animal husbandry operations

— bullfrogs aren't likely to be interested. Their slow growth rate also complicates things, as does their tendency to eat other bullfrogs. Many attempts at raising bullfrogs were short-lived, and the unwanted animals were eventually released into the wild. Most invasive populations today are connected in some way to former frog farms, descended from animals

Since the late 1800s humans have been transporting bullfrogs across the globe in attempts to establish frog farms, where frogs are raised and harvested for their meat. Frog meat, considered a delicacy in Thailand, is being sold at this market in Bangkok.

that managed to escape or were let loose after the exasperated farmers had finally had enough. Sometimes such releases occurred on a large scale; after a government-run frog farm in the Philippines went belly-up during the 1960s, 14,000 bullfrog adults and tadpoles were released into two lakes.

Many frogs would not last long if they were dumped just anywhere. Indeed, increasing numbers of amphibians are disappearing from their native ranges for that very reason — because of habitat loss; they simply can't make it under different environmental conditions. But bullfrogs don't have that problem. They spend almost all their time in the water (unlike many frogs, which as adults live on land), preferring relatively deep, stable, non-moving aquatic habitats. As it happens, such characteristics define many bodies of water in human-modified environments: reservoirs, farm ponds, irrigation channels and other waterways that have been altered for flood control, even garden water features. While the preferred habitats of most frog species have dwindled, bullfrogs have seen theirs undergo large-scale expansion.

Bullfrogs are also highly adaptable. In their native range they can survive in conditions that range from cold Ontario winters (which they survive by hibernation) to the extreme heat of the southern United States and northern Mexico. This ability to deal with climates ranging from subarctic to subtropical is just one of several reasons why they're flexible enough to thrive and build invasive populations across a wide geographical area. Another factor is a slow-but-steady approach to reproduction. Unlike most frogs, bullfrogs often spend two full years as tadpoles, and sexual maturation is also a relatively slow process. While some types of frogs are laying eggs within months of hatching, bullfrogs don't get going until they're from three to five years old. When eventually they do get underway, however, they roll like big trucks. Small female bullfrogs can produce between 6,000 and 7,000 eggs at a time, and as they get bigger they can lay up to 25,000 eggs per clutch. What's more, the females produce eggs continually, meaning that under certain circumstances they can lay more than once a year. By comparison, none of the frogs native to western North America ever generates more than 5,000 eggs a year.

Bullfrogs' survival is further ensured by a formidable set of predatory skills and a varied diet. As carnivorous amphibians, they prey on an impressive list of other species, including fish, crayfish, water beetles, snails, dragonfly larvae, turtles, bats, voles and other small mammals, ducklings, snakes, lizards, salamanders and other frogs. But it's their capacity for stuffing things into those gaping mouths that's most impressive. In 2002, researchers studying ducks had affixed radio transmitters to a young family. When they found themselves a duckling short, they tracked its signal zooming around the pond, apparently underwater. It wasn't until the next day that they found the duckling inside a dead bullfrog that had a transmitter antenna protruding from its gut. Even more amazing, the predatory frog weighed less than 3 ounces (84 g) — it was a mere youngster.

An ability to exploit varied food resources is impressive, involving not just the physiological tools required to eat and digest a lot of different foods but also the sensory skills and dexterity to catch prey that live in different places and move in different ways. The benefits are twofold: bullfrogs remain comparatively unscathed by fluctuations in any one type of prey, and they're capable of finding food even in a foreign environment filled with exotic animals. And what about the cannibalism? Scientists have long been aware that bullfrogs are among the species that eat their own kind. Lately, however, there's mounting evidence that such behavior intensifies when bullfrogs invade new territory; researchers examining their stomach contents have routinely come across the remains of other bullfrogs.

Some scientists think that cannibalism may be particularly advantageous for species in which the adult and juvenile forms eat different food, as is the case with frogs. Usually tadpoles and adult frogs have plenty to eat because their diets consist of rapidly reproducing food such as algae and insects. If for some reason the eggs and tadpoles die off en masse one year, the frog population isn't in much danger as long as there are adults around to lay more eggs. If there's a die-off due to starvation among adult frogs, however — often a possibility, especially in hostile environments — it's a more serious problem, because several years must pass before the population can regain its pool of reproducing adults. Cannibalism thus acts as a form of insurance. Whenever adults find themselves with prey shortages, their ability to eat young frogs allows the population to continue producing large numbers of young. Also, by eating their tadpoles and young frogs, adult bullfrogs are indirectly tapping into food sources that they can't exploit directly.

Although it's difficult to say which traits have contributed most to the bullfrog's success, the success itself has become impossible to ignore. In Arizona, for instance, researchers have reported densities of around 400 bullfrogs per acre (1,000 per hectare) — some 20 to 30 times greater than normally found in the species' native range. On Vancouver Island, bullfrogs have spread as far north as Nanaimo, where residents reported being inundated during the summer of 2008. In Europe in the late 1990s the species was confined to Italy, the Netherlands and France. When researchers conducted a census in 2005 and 2006 in southwestern France and across the Continent, they discovered that bullfrog populations had risen dramatically during the past decade. In southwestern France the frogs were widespread, inhabiting 123 different wetlands. Elsewhere in Europe they had been introduced on at least 25 occasions in eight different countries and were now established in four: Greece, Germany, Belgium and Italy. "Our study shows an alarming increase of the presence of *R. catesbeiana* in Europe, probably caused by the combined effect of multiple introductions from North America, secondary translocations within European countries, and natural expansion." Other areas that now have established populations include Mexico, South America, several Caribbean islands, China, Japan and Southeast Asia.

The spread of bullfrogs arouses mixed feelings. On the one hand, efforts to establish an alternative source of protein for human consumption have paid off. The only difference is that instead of the frog legs coming exclusively from farms, they're now also available in the wild. The downside, of course, is the environmental impact. For decades it has been widely believed that bullfrogs compete with and prey upon native frogs. More recently this pressure has come to be regarded as a major factor contributing to the worldwide decline of amphibians. Recent studies suggest that bullfrogs may be playing a role in the spread of *Batrachochytrium dendrobatidis*, the deadly chytrid fungus. For some reason bullfrogs seem to be susceptible to infection by the fungus but don't usually develop the symptoms that have decimated frog populations around the world. There is now growing suspicion that invasive bullfrogs may be contributing to the epidemic. In 2006 an international team of researchers collected tissue samples from otherwise healthy American bullfrogs in Canada, the U.S., Brazil, Uruguay, Great Britain, France, Italy and Japan. The chytrid fungus turned up on frogs from every country except Japan. (See chapter 3 for more on this deadly fungus.)

In the past, attempts to deal with invasive bullfrogs have involved considerable effort and expense. Shooting adult frogs one by one is usually part of the approach. Another is draining a body of water and then killing the frogs and tadpoles left behind. One such eradication campaign took place in England during the late 1990s. Stocking of private ponds with bullfrogs throughout the country during the 1970s and '80s had resulted in their spread to native ponds in the southeast. When a breeding population was discovered in the county of Kent, conservationists mounted a campaign that involved surrounding the ponds with frog-proof plastic fences, draining the water and then dredging the exposed bottom. This resulted in a cull of an estimated 12,000 bullfrogs and bullfrog tadpoles over a five-year period, and in 2004 the program was declared a success. For local conservationists, however, the presence of other bullfrogs throughout England means there's always a chance the ponds will be reinvaded by a new wave of aliens.

One challenge in the war on bullfrogs is that these hardy amphibians are much more comfortable living around humans than are most native frogs. A recent study in New Jersey's Pine Barrens wilderness area found that wetlands that had been impacted by human activity — impoundments, abandoned bogs, waterways near agricultural lands and residential developments — usually contained only bullfrogs. At the same time, water bodies still in their natural state were found to be largely bullfrog-free. Researchers in California studying wetlands south of San Francisco discovered the same thing. They compared the characteristics of habitats where nonnative bullfrogs were present with those supporting dwindling populations of native frogs. Their results show that the presence of bull-frogs is overwhelmingly associated with human-induced modifications to the environment.

It's not easy being green: for decades conservationists have blamed the bullfrog for the demise of native frog populations, but other factors — like habitat loss — are also responsible.

A good example of how the human world has tilted the tables in favor of bullfrogs is the endangered red-legged frog, the largest native frog species west of the Rocky Mountains. These frogs (probably the ones Mark Twain wrote about in "The Celebrated Jumping Frog of Calaveras County") first ran into trouble in the late 1800s, when they were featured on the menus of restaurants that sprang up during the gold rush. In spite of good evidence that overharvesting had decimated the populations — up to 80,000 red-legged frogs were harvested annually - the subsequent spread of bullfrogs made these aliens convenient scapegoats for the native frog's continued decline.

It now seems more likely that the red-legged frogs were victims of a near-perfect storm of environmental abuse. The first hit was habitat destruction. As humans moved into California, they caused, as in many other places, widespread transformation of freshwater habitats. The rivers and streams that flowed from the mountains into the Pacific Ocean had been dynamic environments marked by seasonal flooding, strong currents and, in the case of the smaller streams, lakes and ponds, partial or complete drying up during the summer. As humans inhabited this landscape they changed the waterways, affecting how rivers and streams meandered and created sandbars, when they flooded, how fast the water moved and how warm it got at different times of the year, what types of vegetation grew in and near them, what materials washed from their shores, and so on. Many large, shallow wetlands that typically dried up in the summer were converted into smaller, deeper ponds that held water year-round. Such changes are thought to have eliminated many of the conditions to which native frogs were adapted, while at the same time creating habitat that's ideal for bullfrogs. For example, frog species whose tadpoles hatch and grow in a few months can survive in a wider range of wetlands, even temporary puddles that hold water for only part of the year. On the other hand, bullfrog tadpoles require permanent deep water in order to complete their two-year maturation cycle.

In part because of the plight of its native frogs, California has invested a considerable sum in trying to understand the detailed ecology of bullfrog–native frog interactions. The result has been several studies that show how habitat modification can create effects that ripple throughout the aquatic ecosystem. Hydrology is certainly a big part of it. Bullfrogs in California are rare in places where streams and rivers have retained their native flow patterns, but common where humans have created deep, permanent water bodies. And researchers have also found that while native frogs appear to be sensitive to water quality, including changes in levels of phosphates, ammonia and salinity, bullfrogs seem less sensitive. They can tolerate a wide range of conditions, including, as one report recently concluded, "highly compromised water quality."

By changing the physical parameters of a freshwater wetland, humans also change the playing field for all life-forms in the ecosystem, and this results in a cascade of ecological readjustments. Consider, for example, aquatic vegetation. In shallower untamed waterways researchers have found that plants tend to be widely distributed. In managed waterways, however, the vegetation is more concentrated, growing in isolated clumps and in a narrower band along the shoreline. Because vegetation is a focal point for insects and other aquatic life, such changes have important repercussions. Studies have shown that red-legged frogs are able to share a habitat with bullfrogs if the aquatic plants are scattered, possibly because it decreases their risk of being eaten while foraging. Where vegetation is clumped, all the food is in fewer places, and so are all of the bullfrogs.

California's wetlands have undergone another major human-driven change: the introduction of nonnative fish. Originally most of the region's lakes were completely free of fish. However, during the latter decades of the 19th century alone, at least 35 different species were introduced to establish sport and commercial fisheries. Several of these introductions — trout, bass and bluegills, to name three — were nonnative predators whose tastes include frogs, tadpoles, frog eggs and sometimes all of the above. Today many of these fish species are well established, and there's mounting evidence that they have made things worse for the region's native frogs. On the one hand, certain fish appear to find bullfrogs and their eggs and tadpoles unpalatable. At the same time there's evidence that when such fish are present, they make life easier for bullfrogs by gobbling up potential competitors. In one study researchers raised bullfrog tadpoles in meshed enclosures that had been placed in ponds in Oregon's Willamette Valley. In some of the enclosures the scientists added native dragonfly larvae; in others they combined the tadpoles with dragonfly larvae and bluegills. When sharing their space with native dragonflies, only 10 percent of the bullfrog tadpoles survived. When the fish were present, the bullfrog survival rate rose to nearly 60 percent.

Such studies reveal how changing the nature of a habitat can unleash complex ecological effects. But the insights gained from this research may be helping change the way in which we deal with problems caused by out-of-control invasive species. Traditionally, wildlife managers in California have dealt with bullfrogs as they do in most other places, relying on a combination of shooting and regular drainage of ponds. In recent years, however, it has been recognized that if you want to save native frogs you also have to save their habitat. Conservation programs have begun to supplement bullfrog-eradication campaigns with habitat restoration. This includes creating gravel beds (favored by certain native frogs) and subjecting wetlands to periodic water fluctuations to mimic natural conditions. Unfortunately, turning back the environmental clock doesn't come cheaply: the U.S. Fish and Wildlife Service has estimated that complete recovery of the red-legged frog will cost at least $10 million.

BROWN TREE SNAKE

Boiga irregularis

I F THERE'S A SINGLE MOMENT that defines the threat of invasive species more than any other, it may be the one experienced by Julie Savidge while on the island of Guam. In the early 1980s Savidge was a graduate student in ecology; she had come to this corner of the South Pacific to study the mysterious disappearance of the island's native birds. At the time some people suspected that disease was to blame, and to test this hypothesis Savidge had established a colony of bridled white-eyes in her temporary lab. White-eyes are dainty green and yellow wren-sized forest birds. Like several other local species, they had become increasingly rare in recent years — so rare that Savidge had been forced to stock her lab with birds from a nearby island.

If an epidemic was to blame for the disappearance of Guam's white-eyes, Savidge wasn't able to find any evidence. What she did discover was how woefully defenseless these tiny birds seemed to be. This was particularly true at night, when they would line up shoulder-to-shoulder on a branch and settle in for a deep sleep that was hard to disturb. Savidge could pick up a bridled white-eye in her hand when it was in this state, and still it would continue to sleep. This trait became more than a curiosity when she walked into her lab one morning in 1982 and discovered carnage. A brown tree snake, apparently entering the lab through an air conditioning vent, had swallowed three of the white-eyes and killed a fourth. From the evidence it seemed obvious what had transpired: the predator had slithered along a branch and gobbled down one bird after another while the others continued to sleep, oblivious to what must have been at least some degree of mayhem.

In the years that followed, Savidge managed to convince the world that brown tree snakes were responsible for what still stands (if you're not counting humans) as one of the worst paths of destruction ever carved out by a single invasive species. Introduced to the island sometime after the Second World War, these sneaky predators are now widely blamed for having driven many of Guam's native vertebrates into extinction. This includes most of the island's native birds, two of its three species of bats (the island's only native mammals) and six of its twelve native lizards. Other species — five of the remaining lizards, for example — have been drastically reduced in number. Meanwhile, the impact has been linked to broader ecological upheavals, including an explosion of spiders and the decline of trees that formerly depended on native birds for pollination. And that doesn't even touch on the power outages, or what appears to be the snake's penchant for taking bites out of young children.

Brown tree snakes (*Boiga irregularis*) are native to the northern and eastern coasts of Australia as well as Papua New Guinea and several other islands of Melanesia. They normally grow to between 3 and 6 feet (1–2 m) in length and can be recognized by their large head, bulging eyes and slender body. Usually they're light brown with dark markings, although in some places they can take on a more striking appearance. In parts of northern Australia, for instance, they're known as "night tigers" because their distinctive body pattern features alternating bands of creamy white and reddish brown. Brown tree snakes are nocturnal and largely arboreal, although they're also comfortable on the ground. They're sometimes described as having bad tempers because of their willingness to strike repeatedly when disturbed. They're also poisonous. Fortunately their fangs are situated toward the back of the jaw, which usually makes it difficult to inject venom into larger objects, including human forearms.

Unlike invasions by larger predators such as mongooses and foxes, the introduction of brown tree snakes into Guam is believed to have been unintentional. Although no one knows for sure, it is widely thought that the snakes were accidentally imported by the American military during restructuring of its network of bases at the end of the Second World War. One clue is that they were first seen in the early 1950s near Apra Harbor, the island's busiest seaport. Also, the snakes on Guam closely resemble those found in the Admiralty Islands, to the north of Papua New Guinea. These islands are not only part of the snake's native range but also a point of origin for military equipment shipped to Guam in the late 1940s. Like all snakes, brown tree snakes are well-known for their ability to survive for long periods without a meal. This, coupled with the fact that they're active at night, has led to the suspicion that they arrived on Guam as hitchhikers, hidden inside cargo. And given that some female snakes can store sperm and lay fertilized eggs years after mating, it is theoretically possible that a single snake was all it took to start a new population.

Once on the island, the species found an inviting home. Picture Guam as a 30-mile-long north/south peanut; Apra Harbor is two-thirds of the way down the western coastline. From their initial invasion point, the snakes began fanning out south, east and north during the early 1960s. By the middle of that decade they had infiltrated the entire southern end of the island. By the late 1970s large numbers had advanced into the entire northern half, except for a couple of isolated patches. By the mid-1980s they were common throughout the entire island. While its spread wasn't particularly fast, other aspects of the invasion illustrate its extent. Researchers were finding brown tree snakes on Guam that were up to 9 feet (3 m) long — 50 percent longer than the species' usual size in its native range. The snakes on Guam also seemed to be occurring in unusually large numbers. When the first serious census attempt was made in 1985, researchers discovered areas in northern parts of the island with up to 100 snakes per hectare — densities unprecedented in the snake world.

Because brown tree snakes remain well hidden during the day, Guam's residents didn't have a clear picture of what was going on in the forests and fields around them. However, there were some very obvious hints.

The brown tree snake has become a major pest in Guam; ▶
these snakes were electrocuted after getting too close to a power
box. The snakes are responsible for hundreds of power outages
— and millions of dollars in associated repairs — each year.

INSERT PAYMENTS HERE
DROP BOX
GUAM POWER AUTHORITY
GPA
POWER TO SERVE

In developed areas the snakes were mistaking power-line poles for trees and climbing along the wires in search of roosting prey, dragging them down under their weight. According to the U.S. Geological Survey the snakes were responsible for 1,600 outages between 1978 and 1997, with the associated repair bills adding up to several million dollars a year. Even more unsettling, medical authorities began receiving an increasing number of mysterious snake-attack reports from individuals claiming to have been bitten while they slept. On more than one occasion the snakes seemed to zero in on sleeping infants and small children. What's more, the attacks included not just multiple bites but also apparent attempts at constriction. Eventually it was realized that these weren't disoriented snakes striking out in self-defense, but rather active hunters on the attack.

It was the mass extinctions, however, that attracted the most widespread attention. Because people rarely see brown tree snakes eating anything in the wild, pinning the loss of Guam's native species on these predators hasn't been easy. Julie Savidge's approach was to compile government wildlife records, news reports and records of snake-related power outages. She combined these data with interviews of more than 350 Guamanians in order to create detailed chronologies of the rise of the snakes and the decline of the island's forest birds. The results were dramatic. During the 1960s, bird species began disappearing from the south. As the snake population spread northward during the 1970s, birds started becoming rare in that part of the island as well. By the early 1980s, the only place where all 10 of the forest species Savidge studied could still be found was one of two locales that had yet to be overrun by snakes. By the late 1980s, those species could no longer be found anywhere on the island; the list included white-throated ground-doves, Mariana fruit-doves, Micronesian kingfishers, rufous fantails, Guam flycatchers, bridled white-eyes, Guam rails and Micronesian honeyeaters. Two additional species — Micronesian starlings and Mariana crows — had been reduced to fewer than 100 individuals. And those were just the forest birds. At around the same time the island saw the loss of cliff-nesting seabirds such as the white-tailed tropicbird, the fairy tern, the brown noddy and shorebirds such as the Pacific reef heron.

Since Savidge's initial studies, other researchers have implicated brown tree snakes in the disappearance of several additional species, including sheath-tailed bats, little Mariana fruit bats, snake-eyed skinks and spotted-belly geckos. By eating all the insect-eating birds, the snakes also stand accused of having triggered insect outbreaks that have resulted in the spread of disease, defoliation of trees and dramatic increases in the island's spider populations. Evidence that the invasion may be having an even broader impact came in 2008, when a team of Danish biologists released the results of a study linking the rise of brown tree snakes with what appears to be reduced reproductive success among native trees. The researchers examined two tree species — a mangrove tree and a high-canopy forest tree — and compared pollination success rates on Guam with those on the nearby island of Saipan. They discovered that the trees on Saipan were visited almost exclusively by native birds, and that they also set fruits at a far greater frequency than on Guam, where the flowers were visited only by insects, never birds.

While the brown tree snake has become one of the defining cases of invasive-species biology — a classic example of what happens when a predator is unleashed on an island filled with unsuspecting prey — some mystery still surrounds exactly how a single species could have caused so much damage. It's easy to imagine the snakes eating their way through prey that lacked any instincts for fleeing the danger posed by predators, but it's somewhat harder to imagine how this slaughter was able to continue once the individual birds became rare.

Brown tree snakes are impressive hunters. Research on Guam indicates that the species uses two completely different approaches: active foraging, which requires locating potential foods such as eggs mainly by smell, and a more vision-oriented ambush approach, sneaking up on and suddenly striking more agile prey such as full-grown birds and lizards. Researchers have also discovered that brown tree snakes avoid certain scented lures, which also suggests that their hunting is driven by a sophisticated combination of both night vision and smell.

Brown tree snakes also appear to be willing to eat things they've never seen before. In addition to birds, eggs, lizards, bats and rodents, they've been seen swallowing dog food, chicken bones, rotting animal carcasses, nuts and even bits of human garbage. If they have any problem when it

The brown tree snake is known for its bad temper, and will strike repeatedly if agitated. Exceedingly bold, it's been linked with nocturnal attacks on sleeping children.

comes to diet choices, it may be their inability to gauge when a target is out of their league. In the lab they frequently attempt to engulf rodents that are too big to swallow, and it's been speculated that their attacks on infants stem from the same deficiency in judgment. As their presence in humans' warehouses and bedrooms suggests, they're also bold, and keen to explore new types of habitat.

Despite these advantages, scientists have had a hard time trying to understand why the brown tree snake has been such a successful invader in Guam; brown tree snakes seem to lack specific traits that set them apart from other snakes. Indeed, one trio of American reptile experts concluded, "the brown tree snake is not unique in either its ecology or its behavior." Added to this conundrum is the observation that the snakes are not unusually dominant in their native range. Nor have they managed to invade any other new territories, despite good evidence that ships and planes from Guam have been inadvertently transporting them around the world for half a century. The list of places where brown tree snakes have been seen slithering around cargo crates and customs zones includes islands in both the nearby South Pacific and the Indian Ocean, as well as Singapore, Japan, Taiwan, Spain, Texas and Hawaii.

As a cause of the widespread decline of Guam's native species, perhaps the brown tree snake is not entirely to blame. This reptile's notoriety has created an impression that the island's ecological innocence was lost by a single historical event: the fateful day when a supply crate was opened and an invasive species slithered into the virgin wilderness. But nothing could be further from the truth. As in many other islands in the South Pacific, the ecosystems of Guam have been under steady assault ever since humans arrived some 3,500 years ago. The original inhabitants cleared native forests and they're also thought to have introduced many new species, including Asian black rats, monitor lizards, various geckos and mosquitoes. If archeological digs on nearby islands are anything to go by, these changes likely resulted in loss of a large fraction of native species that had been present for millennia.

In the 1500s the pace of change accelerated with the arrival of Spanish conquistadors, who introduced, among other species, Philippine turtle-doves, blue-breasted quail, Philippine sambar deer, Polynesian rats, roof rats, house mice, feral pigs and Asiatic water buffalo. While many of these aliens thrived, often at the expense of native wildlife, the mammals had a particularly devastating impact. Deer and buffalo feasted on the seedlings of native plants that had previously never known browsers, and the feral pigs literally uprooted the forest floor.

And the damage continued. During the 20th century, when Guam was in American hands, native forests were cleared for coconut plantations and wetlands were drained for cultivation of taro and rice and for building port facilities, towns and cities. During the Second World War Guam fell to the Japanese, only to be reclaimed by the Americans in an extended battle that saw much of the island's forests and caves bombed to smithereens. When the dust had settled, the decision to make Guam a strategic military outpost resulted in rapid human population growth and further clearing of land for runways, warehouses, housing, golf courses, sugarcane plantations, urban areas and resorts.

To add insult to injury, after the war and throughout the 1960s, most of Guam — its wetlands and forests and even its caves — was regularly sprayed with insecticides, including DDT, in an effort to control mosquito-borne malaria. And the nonnative species kept on coming: the curious skink, the Indian musk shrew, an arboreal lizard called the green anole, the pheasant-like black francolin, the Norway rat, the Eurasian tree sparrow, feral dogs and cats and a notoriously aggressive forest bird, the black drongo. And that's just the animal invasions. To prevent soil loss on lands denuded by the ravages of war, large parts of Guam were planted with *Leucaena leucocephala*, a small, fast-growing nonnative tree known locally as tangantangan that has taken over much of the island and suppressed regeneration of native plants. Various weeds, among them chain-of-love, dodder, agalondi and mimosa, have also run wild after being introduced, strangling native vegetation in the process.

Clearly Guam's flora and fauna are suffering from more than just brown tree snakes. Indeed, five of the 23 native bird species common after the war — including the Micronesian megapode, wedge-tailed shearwater, Mariana mallard and white-browed crake — disappeared either before the snake invasion occurred or for reasons that had nothing to do with snakes. Two additional species, the brown booby and the nightingale reed warbler, were already declining before the snake invasion, and some scientists have argued that pesticides played a major role in reducing the number of cave-roosting island swiftlets, whose population dwindled first in the north. The little Mariana fruit bat and the sheath-tailed bat were not being seen by 1968 and 1972 respectively, which suggests that they may also have declined in northern Guam before the arrival of the snakes. (Supporting this notion is evidence that the sheath-tailed bat also declined on snake-free islands about the same time it was disappearing from Guam.) Finally, the disappearance of at least three native lizards — the azure-tailed skink, Mariana skink and blue-tailed copper-striped skink — occurred during a rapid expansion of invasive musk shrews, a potential competitor.

This is not to suggest that the brown tree snake is innocent. But what it does suggest is the possibility that the invasion equation wasn't as simple as is widely believed, and it raises some fundamental questions that remain unanswered. For example, would the birds have been able to survive the snakes' presence in the absence of the many other stresses that plagued the island's ecosystems? It's been suggested that one of the reasons the snake was able to eat all the birds was because it was able to maintain high densities even when birds were becoming rare. And one reason for that may have been the island's unusually high densities of so many other forms of nonnative prey, particularly introduced shrews, rats, mice and lizards.

At the same time, one might also wonder what shape Guam's bird species would be in even if snakes hadn't invaded. On nearby Rota, for example, there aren't any snakes but there has been a litany of bad news about native bird species: island swiftlets were extirpated after the war; Rota bridled white-eyes are listed as critically endangered; and a survey published in 2008 showed that seven of eight other forest-bird species experienced sharp population declines between 1982 and 2004, five of them by 60 percent or more. Similarly, attempts to reintroduce captive-bred native rails on Guam failed even though they were conducted in a large habitat encircled by a snake-proof fence; the failure has been attributed to predation by feral cats.

In the end, the full invasive potential of brown tree snakes may never be known, because wildlife officials and conservationists are doing everything in their power to make sure the definitive experiment — establishment of brown tree snakes on other islands — never occurs. On Guam there is an ongoing drive to develop snake-proof barriers that are high and smooth enough to keep snakes confined, strong enough to withstand cyclone-force winds, and at the same time affordable. Other control approaches that have been considered include direct trapping and killing of snakes, development of snake attractants, release of snake parasites and diseases, and spreading chemicals known to interrupt snake reproduction. Recently U.S. government researchers have been experimenting with dead mice stuffed with acetaminophen, an active compound in human pain-relief medications that is toxic to brown tree snakes. In a test run in 2007, more than 1,100 poisoned rodents were airdropped over Guam's forests. To make sure they got snagged in the canopy, where snakes would be sure to find them, each dead mouse was outfitted with its own pint-sized biodegradable parachute.

Nature is helping out too. The brown tree snake population on Guam appears to have stabilized at roughly 50 percent of their mid-1980s peak. This may be because previously out-of-control populations of introduced species such as musk shrews and curious skinks have now thinned, a development that suggests Guam now has a somewhat stable, albeit completely revamped, food web. At the same time, efforts are being made to prevent further introductions. Authorities have recently turned to tactics such as snake-sniffing dogs and fumigants to ensure that outgoing cargo remains free of live snakes. Snakes seen in foreign ports, needless to say, are destroyed on sight. So far, with the possible exception of Saipan, the spread of the brown tree snake — if not its notoriety — has been contained.

CHYTRID FUNGUS

Batrachochytrium dendrobatidis

IN THE FALL OF 1991, a group of arroyo toads housed at the University of California in Santa Barbara began mysteriously dropping dead. Three of the dead animals, their bodies fixed in formaldehyde, were sent to the U.S. National Institutes of Health in Bethesda, Maryland, where veterinary pathologist Donald Nichols attempted to determine the cause of death. The animals' organs were intact and, apart from what appeared to be a mild skin condition, they seemed in remarkably good shape, despite being dead. Nichols was able to identify some odd single-celled organisms associated with the skin lesions, but he had no idea what they were. His best guess was that they had affected the vital membrane passageways of the skin, somehow suffocating the animals. Back in Santa Barbara, meanwhile, the unidentified disease continued its path of destruction, eventually killing the entire colony.

Around the same time, some 9,400 miles (15,127 km) away in the rain forests of northern Australia, Keith McDonald was struggling with a mystery of his own. Since the late 1970s frogs had been dying dramatically throughout Queensland, in sudden episodes of mass death that claimed whole populations and sometimes even entire species. The declines were first seen around Brisbane. By the early 1990s they had reached the northeastern corner of the state, where McDonald worked as a government wildlife specialist. Something in the environment seemed to be going haywire, killing animals that had existed on Earth largely unchanged for 160 million years. McDonald couldn't figure out what it was.

Meanwhile, American researcher Karen Lips was facing a similar challenge in the high-elevation tropical forests of Costa Rica, where she had been studying stream-breeding frogs since 1990. After coming across several dead frogs in the early years of her work, the young biologist was surprised to discover in 1994 that previously widespread harlequin frogs were no longer common, which she attributed to a delayed rainy season. However, on her next trip, in 1996, she recorded at least 90 percent declines in five different species. That same year she traveled to a site in Panama where she had previously recorded 55 different species; this time she found just 24. Unlike the earlier declines, however, she stumbled upon this one as it was unfolding: dead and dying frogs lay everywhere. After gathering up as many of the victims as she could, Lips had them sent to the Animal Health Diagnostic Lab in Frederick, Maryland.

Back in Australia, researchers began to suspect that a killer virus was on the loose, and Lips thought the same thing, but no one had any luck finding such a virus. What they did find, however, were the same unusual microbes that Donald Nichols had observed earlier. It wasn't until around 1997 that all three groups of scientists finally linked the weird single-celled organisms to a phylum of fungi known as chytrids. The previously unknown fungus was named *Batrachochytrium dendrobatidis*, or *Bd* for short. When news of the discovery spread, many found it hard to believe. Chytrid fungi are best known for decomposing organic matter, and no member of this group had ever been linked before with vertebrate disease. Even more implausible was the apparent magnitude of the epidemic. If microscopic organisms really were killing off frog species on three continents, it meant that scientists had uncovered a disease outbreak whose scale was unprecedented.

Despite early skepticism, *Bd* is now widely viewed as one of the major culprits in recent declines among amphibians worldwide. By 2010 the fungus had infected more than 350 species of frogs and salamanders in 48 countries from every continent except Antarctica. Additional research has identified *Bd* as a factor in the decline of more than 200 species. In Australia alone it is believed to have severely impacted 14 species, including eight that are now likely extinct.

Although it remains unclear exactly how the fungus kills, it is believed to survive by eating keratin, a common protein found in many life-forms and an important building block in the skin tissue of amphibians. In lab studies, healthy adult frogs inoculated with *Bd* typically develop symptoms of a disease — known as chytridiomycosis — that invariably leads to death in less than three weeks. Infected tadpoles don't die, probably because they possess very little keratin. However, these young frogs can carry the fungus for months, and once they pass through the final stages of metamorphosis and their bodies begin making keratin, they get sick and eventually die as well.

Not every amphibian population is equally affected. Northern populations of mountain yellow-legged frogs in California, for instance, seem to suffer less from *Bd* because of the presence of anti-chytrid chemicals produced by bacteria residing on their skins. Many green and golden bell frogs have been eliminated throughout their range in southeastern Australia, but the survivors include populations that inhabit highly contaminated environments such as former gold and copper smelting operations, and it's been speculated that some pollutant may be providing protection. Other species have been hit hard by the fungus but have managed to survive. Studies of these cases have found that the fungus is still present afterwards, at times even widespread, among healthy survivors, providing textbook examples of how sometimes only the fit survive. In still other places, entire species seem to have been wiped off the map.

Infectious-disease experts remain unsure how a microbe could kill every last one of its hosts — a situation unparalleled in the history of disease. One explanation is that a frog species can sometimes amount to no more than a few small populations scattered across a single rain-forest valley. When a deadly disease strikes such vulnerable species, perhaps there is just no escape.

The identification of the chytrid fungus comes amid efforts to understand the much larger decline of amphibians worldwide. According to the latest statistics compiled by the International Union for Conservation of Nature in 2008, some 168 species are believed to have gone extinct in the past two decades, while close to 2,500 other species have been in decline. What's made the phenomenon so puzzling is that many declines have occurred in pristine wilderness areas, far from human interference, and sometimes within only a few months. One well-known example is the golden toad of Costa Rica, a small, beautifully colored species whose range was once limited to less than 12 square miles (30 sq km) of high-elevation rain forest. Beginning in the early 1970s, scientists discovered that hordes of these delicate creatures would gather on the forest floor to mate in temporary pools formed by seasonal rains each April. As many as 1,500 frogs were sometimes counted in the same place at the same time, a spectacle that occurred year after year until 1989, when only a few individuals turned up

African clawed frogs can carry and spread chytrid fungus but never develop disease symptoms. Researchers speculate that these frogs, which were widely exported as subjects for science experiments, pregnancy tests and exotic pets, may be largely responsible for the rapid spread of the fungus around the world.

at the main gathering site. The following year there were no signs of the toad anywhere, and the species hasn't been seen since.

Many amphibian declines have occurred for obvious reasons. According to one recent study, 183 of the 435 species deemed to be "rapidly declining" were found to be in trouble due to habitat loss. Another 50 were determined to be under threat due to over-exploitation (a category that includes frogs harvested for food and frogs, like the poison dart frog, collected for the pet trade). As with the golden toad, however, the reasons for the decline of many amphibians in otherwise pristine wilderness areas are much tougher to understand.

The search for answers began with a basic reality of amphibian physiology: frogs, salamanders and their kin are sensitive creatures. For one thing, they're covered with a thin, porous outer membrane that allows them to breathe oxygen and drink water through their skin. Their eggs lack protective shells, being nothing more than thin-layered, jelly-like bubbles. Lab studies have shown that these features make amphibians highly susceptible to pollutants such as pesticides and other chemicals, as well as to changes in the environment such as increased ultraviolet (UV) radiation or fluctuations in humidity and temperature. Research in the field has found evidence that pesticides, increased UV light and climate change may indeed have contributed to some recent declines.

Then there's *Bd*. Since it became known that a chytrid fungus was lurking at the scene of die-offs on three continents, researchers have been trying to unravel the many mysteries surrounding this epidemic. Where, for example, did this brutal serial killer come from? No one knows for sure, and there are two opposing theories. It could be that the severity of the disease is what's new, not the fungus itself. According to this idea, *Bd* may have been for the most part an innocent bystander, an innocuous and possibly widely distributed part of the environment whose anonymity was ensured by its limited success. But then something may have changed in the environment that effectively altered the fungus's standing in the ecological order of things, that made it easier for it to thrive. Or environmental change may have rendered amphibians more susceptible to infection, possibly by weakening their immune systems or disturbing the resident microbes that higher life-forms depend on for defense against infectious disease.

Some climate experts have argued that global warming facilitated the spread of *Bd* among frog species that declined recently in Central America. Studies from Italy, meanwhile, have revealed that *Bd* was widespread among local frogs as far back as 1999; despite annual sampling, researchers detected no signs of chytridiomycosis until the summer of 2003, when southern Europe experienced record-breaking heat.

Another theory about the origins of the epidemic views *Bd* as an emerging infectious agent that has spread around the world only during recent decades. According to this idea the fungus has dispersed from its ancestral point of origin and in the process come into contact with new species that have had no opportunity to evolve defenses. In this regard, *Bd* is to virgin amphibian populations what smallpox was to the indigenous peoples of the New World. Backing up this idea are recent genetic tests that show a high degree of similarity among *Bd* samples collected from all over the world, evidence that *Bd* isolates from as far apart as Australia and Panama likely shared a common ancestor sometime in the not-too-distant past.

The emerging-epidemic idea is further supported by ongoing detective work by frog scientists around the world ever since the fungus was identified as a threat. By testing for *Bd* among frog samples in museums collected during the past century, researchers now have a rough idea of when the fungus reached specific areas before its association with the declines was known. Monitoring wild populations during the past two decades has also allowed scientists to track the killer's initial appearance. The results show a consistent worldwide pattern in which the first signs of the fungus are quickly followed by the first known amphibian declines. The pattern first appears in the United States around 1974, in northeastern Australia during the late 1970s and in western Australia in 1985, and finally in Costa Rica and Venezuela around 1986.

The most detailed records come from work done by Karen Lips and

Corroboree frogs are among the 14 species of amphibians in Australia ▶ to have been severely impacted by chytrid fungus. These Corroboree eggs are being kept at the Taronga Zoo in Sydney, Australia, to help replenish the species, which is nearly extinct.

her colleagues in Central America. According to their sampling studies, *Bd* spread southward through Costa Rica at the rate of 16 miles (26 km) per year, reaching Panama in the early 1990s; by 2008 it had crossed the Panama Canal. At the same time the researchers recorded frog declines occurring in the wake of the fungus's spread. Frogs at El Copé, Panama, were free of *Bd* until the fungus showed up in late September of 2004. By mid-January of the following year, the area's amphibian population had been reduced by half, with 38 species affected; tests revealed that almost every dead frog was infected with *Bd*.

The earliest signs of *Bd* that researchers have uncovered come from museum samples of African clawed frogs that were collected from South Africa in 1938. In the wild these frogs have been found to carry the fungus without any signs of disease, which suggests a long period of co-evolution and points toward Africa as *Bd*'s point of origin. Additional proof comes from the fact that African clawed frogs have been widely exported during the past half-century, initially as a primitive human pregnancy test (a female frog will ovulate when injected with human urine containing pregnancy hormones) and later for widespread use in scientific research and as exotic pets.

Although there's no clear proof that trafficking in clawed frogs spread the fungus, the history of the trade would explain how it managed to spread globally in such a short time. Local dispersal, meanwhile, may have been facilitated by a combination of factors, including the ability of tadpoles to remain healthy even while infected (thus serving as both incubators and delivery vehicles for the fungus), the fact that the fungus is often found in water (and thus gets swept along by currents), and the hardiness of the fungus spores (they may have been moved from one exotic location to another on the tires of a vehicle or the boots of an unwitting hiker or scientist). Another, recently discovered culprit is the American bullfrog, a widely distributed invasive species that appears to be able to carry the fungus without getting sick.

Trying to understand the full extent of the epidemic has proven to be another challenge. One reason is that die-offs in remote areas often occur without leaving behind any clues — frogs may die and decompose before anyone has a chance to find them. Scientists are left with only a few indirect pieces of circumstantial evidence to go on. In one analysis, nearly half of all declines — 202 of the 435 cases — were deemed to be "enigmatic." Recently, however, researchers in Australia have argued that *Bd* is the only plausible explanation for most if not all of the declines that can't be explained by habitat loss or over-exploitation. For one thing, the speed with which the declines occurred suggests that something is killing adult frogs. And the only possible explanations for mass death of mature frogs, they argue, are poisoning, some form of environmental catastrophe, or disease. In the pristine areas where most of the unexplained declines have taken place, there are no signs of any large-scale poisoning or habitat change. That leaves *Bd*.

Even if these researchers are correct, there remains the question that so often accompanies sudden species invasions: is this an opportunistic scavenger taking advantage of hosts weakened by other causes, or is it a cold-blooded serial killer wiping out populations that are otherwise healthy and stable? The answer may be a little of both. Often the declines affect populations that show no signs of suffering from anything other than infection. Perhaps not coincidently, such declines have occurred most frequently in areas where the environmental conditions match those identified as ideal for the growth and survival of *Bd*. In the rain forests of northern Australia, for example, this seems to be stream habitats at higher elevations. Most of Central America also appears to be a prime breeding ground.

But the outcome of a *Bd* outbreak might be amplified by the depleted state of its host. Take the Kihansi spray toad, for example. Found only in a few mist-sprayed pools near the waterfalls of Tanzania's Kihansi River, the toads declined dramatically following a hydroelectric project that diverted much of the river's flow away from a series of waterfalls. In their desperation to save the species, conservationists went so far as to install sprinkler systems to mimic the spray that had previously been a feature of the habitat. For a while it appeared to be working — the estimated number of toads had risen dramatically to 20,000 by June 2003. But then things went quickly downhill. Six months later, researchers found only five toads. Disease is likely to have played a role; it was around this time that the scientists first saw signs of *Bd*. But the crash also occurred not long after the river dams had been opened briefly to flush sediments that had accumulated in upstream reservoirs. These sediments were laden with pesticides and fertilizer from surrounding agricultural lands, suggesting that pollution may have played a role as well.

Some observers think both of these factors were only secondary to the main problem of habitat loss. For one thing, the sprinklers were never able to completely mimic the natural spray of a waterfall, which had carried not just water but also silt that would have nourished nearby plants, insects and microbes. As a result, the ecosystem within the spray zone changed completely, with different plants and insects replacing the native life-forms. Although it's likely no one will ever be able to say so for sure what happened, the feeling is that the toads were already vulnerable, weakened by the effort of trying to survive in a world that no longer bore any resemblance to the one in which they had evolved. On this uneven playing field, *Bd* was able to move in for the kill.

In an attempt to limit further losses, scientists worldwide have begun removing frogs from areas thought to lie in the path of *Bd*'s ongoing wave of destruction. The hope is to establish captive breeding populations that may be able to replenish these species should they get wiped out in the future. Such measures aside, it appears that little can be done except watch and learn. Scientists involved in the case argue that continued research is necessary in order to better understand the factors that led to such a destructive natural force being unleashed so quickly. The unspoken and somewhat unsettling message is that if it can happen to the frogs, it can happen to any form of life, including humans.

HUMBOLDT SQUID

Dosidicus gigas

FOR THE OCEANOGRAPHERS working aboard the research vessel *Maurice Ewing* one mid-September night in 2004, it must have seemed like something right out of a Jules Verne novel. While drilling for core samples from the ocean floor off Alaska, the scientists noticed strange shadows darting into the large school of baitfish that had been attracted to the ship's deck lights. At first there were just two or three, but as crew members were hauling in a core sample they found themselves staring into a sea that was boiling with activity — about a hundred dark bodies, many at least 5 feet (1.5 m) long, had converged on the baitfish in a ferocious feeding frenzy that lasted for a full hour.

From the tentacles flailing out of the water the researchers knew they were surrounded by a pack of squid, but these were clearly unlike any squid that had been seen before in Alaskan waters. When the researchers managed to spear one, they discovered it was indeed an alien species — the Humboldt squid, a notoriously terrifying sea creature whose nearest rightful home lay more than 2,000 miles (3,200 km) to the south, in the subtropical seas off the coast of northern Mexico.

If this had been an isolated incident, it is likely that what was observed in the Gulf of Alaska that night would have been forgotten by all but those fortunate enough to witness the spectacle. As it turns out, however, it was just one of a rash of similar sightings that together point to one conclusion: Humboldt squid — the creatures that Mexican fisherman refer to as *los diablos rojos*, or red devils — appear to be undergoing a major range

Each of the squid's tentacles is equipped with thousands of suction cups that are lined with razor-sharp hooks. These hooks help the squid to snare its prey, which might be anything from a tiny crustacean to an unlucky scuba diver.

expansion. As unlikely as it sounds, a deep-sea beast that remains largely a mystery to science has become an invasive species.

Historically Humboldt squid have been found off the Pacific coast of the Americas in tropical and subtropical waters between Baja California and northern Chile. Although the odd individual had been known to stray as far north as San Francisco, such sightings were rare. Beginning around the late 1990s, however, large numbers of Humboldt squid began washing up on the beaches of southern California. In Monterey Bay, 125 miles (200 km) south of San Francisco, researchers were able to pinpoint the invasion during an ongoing 16-year project that involved monthly voyages by a remote-controlled deep-sea exploration vessel. The vessel's camera had recorded no signs of Humboldt squid from the program's beginning in 1989 through to 1997, when large numbers suddenly appeared. Although only a few squid were seen in subsequent years, the animals returned in force in 2002 and have remained in Monterey Bay ever since.

During the past five years fishermen and scientists have begun seeing

Humboldt squid for the first time in areas even further to the north. In 2004 the Royal British Columbia Museum began receiving a stream of sightings — the first ever recorded in the waters off the B.C. coast — including one from a local fisherman who had encountered a school of thousands swimming near the surface at night. Sightings like the one by the crew of the *Maurice Ewing* that same year indicate that the squid have moved as far north as the Gulf of Alaska. Since then there have been regular sighting along the west coast each year, including the summer of 2009, when hundreds of dead Humboldt squid washed up on beaches along the Pacific coast of Vancouver Island. Meanwhile, in Chile fishermen have been seeing the species to the south of its normal range.

What's going on? On the one hand, it may be nothing extraordinary. Ocean animals, especially large ones, are often seen in strange places, their normal sense of place being temporarily disrupted by storms or short-term climate changes. But Humboldt squid are thought to live for only a couple of years. This, coupled with their continuous presence for more than seven years in some new areas, suggests that the red devils may be undergoing a long-term range expansion.

Certainly this species possesses many characteristics of a successful ecosystem invader. Scientists have not spent a lot of time studying Humboldt squid, but from what they do know it's clear these animals are not fragile creatures. The squid spend most of their days hunting in the deep ocean. Most of those observed by researchers monitoring video feeds from the Monterey Bay deep-sea dives were at depths of between 650 and 3,000 feet (200–900 m), but some were seen more than 6,500 feet (2,000 m) into the blue. During the night the squid are known to follow prey all the way to the surface, a behavior suggesting that these animals are extremely hardy in terms of their ability to tolerate variations in temperature, oxygen levels and all the other varying conditions between near-surface waters and the deep sea. Indeed, recent radio-tagging studies have shown that Humboldt squid appear to be comfortable hunting in parts of the ocean where the oxygen levels seem too low to support the metabolic activity of such an active hunter. How they perform this physical feat remains a mystery.

Like many invasive species, Humboldt squid have the ability to reproduce quickly, not only churning out lots of offspring but also cycling rapidly through generations. What they also bring to this equation is remarkably rapid growth rates. It's not clear exactly how big these creatures can get (they're not to be confused with the infamous giant squid only recently captured on film for the first time), but it's generally assumed that they're capable of reaching lengths of more than 6.5 feet (2 m) and weights of more than 110 pounds (50 kg). What's remarkable is the fact that they can achieve this size during a lifespan believed to be no more than two years — growth that's been found to be as much as 3/4 of an inch (2 cm) a week. Such rapid reproductive abilities and growth rates are unusual in top predators, and the combination has no doubt aided the species' rapid expansion.

And let's not forget that Humboldt squids are ferocious and voracious predators. Like their fellow cephalopods the cuttlefish, they feed using eight outer arms that are good for holding, plus two inner tentacles used for snaring. There's also a hard, razor-sharp inner beak shaped something like a parrot's and made from a strong, flexible material whose physical properties are so admirable that scientists are trying to deconstruct its molecular structure in hopes of engineering something similar. Finally there are the thousands of suction cups lining each of the squid's outer arms. These suction cups are ringed with tiny hooks like piranha teeth that aid in grasping large prey.

As if this intimidating arsenal weren't enough, Humboldt squid possess a terrifying instinct for traveling and hunting in packs like wolves. These packs are believed to number in the hundreds, occasionally comprising as many as 1,200 individuals. One might consider this overkill, since the bulk of the species' prey consists of creatures that are no match for even a single 6-foot cephalopod, a list that includes krill, crustaceans, other squid and a large variety of small fish species. But pack hunting is believed to provide advantages in corralling and devouring large schools of fish, and it may also come in handy when larger prey comes along. Squid have been reported to attack pretty much anything they encounter, including wounded

Nicknamed the red devil (*diablo rojo*) by ▶
Mexican fishermen, the Humboldt squid is
truly a Super Species of mythic proportions.

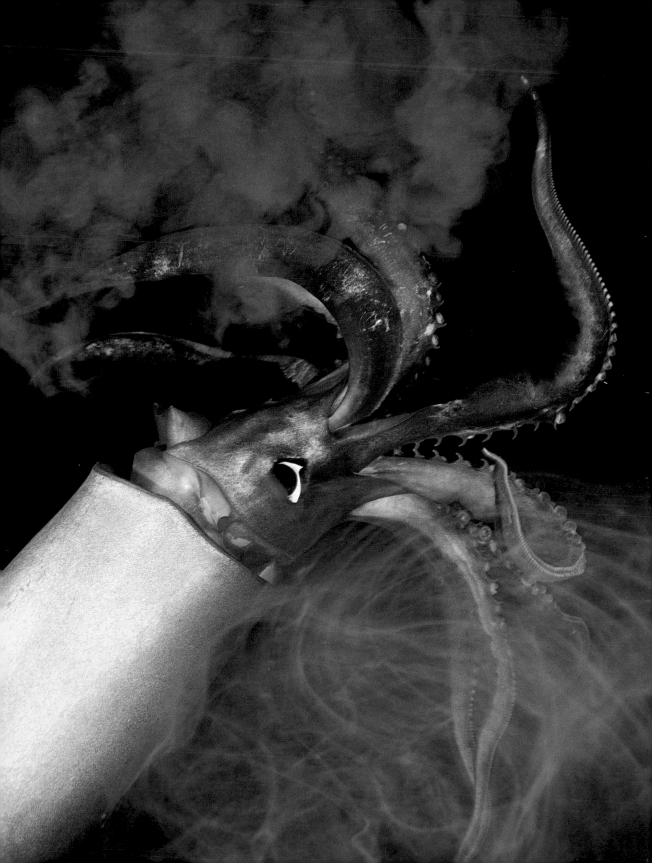

members of their own kind and even occasional scuba divers. Scuba diving videographer Scott Cassell has had numerous close calls while filming Humboldt squid, including one incident when a squid dislocated his right arm and another when a squid dragged him down 30 feet (9 m), rupturing his eardrum. As he once told a reporter from the *Monterey County Weekly,* "Diving with these squid is like diving in a barroom brawl." In Mexico there are stories of fishermen falling overboard and never being seen again. There is some debate, however, over whether attacks on larger prey are a common behavior or anomalous acts committed by rogue individuals.

While it's unlikely that Humboldt squid will ever amount to a serious threat as man-eaters, the species' recent proliferation is of growing concern to both commercial fishermen and ecologists. One worry is that a fast-

Researchers estimate Humboldts can reach lengths of over 6 ½ feet (2 m) and weights of over 110 pounds (50 kg). They tend to hunt in packs of several hundred, and one diver has likened diving with giant squid to "diving in a barroom brawl."

growing predator such as the Humboldt squid may have a major impact on populations of smaller fish. There is evidence that the squids are feasting on lantern fish and other creatures living deep beneath the sea, as well as on hake, anchovies and many other species that are part of the rich aquatic ecosystems in the shallower continental shelf region off the California coast. Already concern is growing that continued expansion of the squid represents a major threat to hake and other important commercial fish in both North America and southern Chile. The monthly surveys in Monterey Bay revealed a sudden drop in hake numbers only months after large numbers of squid made their initial appearance.

Scientists are now trying to figure out what's going on. One possible culprit is warmer water. The summer of 2004 saw the warmest surface temperatures ever for Pacific waters off the coast of North America, and talks with fishermen have caused scientists to suspect that Humboldt squid were present in large numbers during the El Niño years of the late 1990s. According to this line of thinking, the squid may move north with the warm El Niño currents. Temperature may be the enabling factor, but there is also evidence that recent changes have triggered an increase in the flow of subsurface water from the equator toward the poles, as well as in the kinds of plankton favored by larval squid. Both may be additional contributing factors of some importance.

But the picture may be more complicated. For one thing, the apparent persistence of the squid in northern waters in recent years suggests that something in addition to El Niño warming is contributing to the species' proliferation. Some researchers have also pointed out that Humboldt squid are not all that sensitive to water temperature. After all, they spend much of their time hunting thousands of feet below sea level, in water that is often as cold as 46°F (8°C). Now the argument is that something else must be to blame. According to Louis Zeidberg and Bruce Robison of the Monterey Bay Aquarium, that something else is the absence of predators. In a paper published in the *Proceedings of the National Academy of Sciences* in 2007, the two marine biologists suggest that the changes may be related to recent major decreases in the populations of tuna and big game fish, such as swordfish, marlins and sailfish, caused by overfishing. These predators, particularly tuna and billfish, may have restricted the range of Humboldt squid in the past either directly, as predators of juvenile squid, or indirectly, in competition for smaller fish and other prey. Now, with these ferocious fish in rapid decline, the squid are having a field day.

The case of the Humboldt squid is far from closed. Some scientists have pointed out that there isn't a strong correlation between expanding squid populations and the absence of top predators, since dramatic declines in tuna and other billfish were first known to have occurred at least three decades before the first squid expansion. Furthermore, computer models designed to predict the impact of fluctuations in populations of different species on the overall marine food web suggest that the size of the predator pool has little impact on the overall health of the squid population. Rather, squid numbers seem to be driven by how much there is for them to eat. Another possibility is that the squid's spread may be due to a recent sudden and unexplained drop in oxygen levels that has been observed in the near-shore waters off the coast of California. The science of understanding fluctuations of ocean species remains in its infancy. Why these giant squid are on the move — and how far they're destined to go — is a mystery that still remains to be solved.

EUROPEAN GREEN CRAB

Carcinus maenas

WITH THE GROWTH OF HUMAN populations, few aquatic habitats have been subjected to a bigger assault than the sheltered bays and estuaries that lie along continental coastlines. Such areas are ideal for ports, natural ports give rise to settlements, and settlements produce a range of environmental stresses: pollution in the form of runoff, habitat destruction, intensive harvesting of marine resources and nonstop boat traffic, to name a few. The result is ecological chaos for native species and opportunities for invaders — invaders such the European green crab, a feisty opportunist and voracious predator that is quickly taking over coastal habitats from southern Australia to Newfoundland.

The green crab is native to the Atlantic coastlines of Europe, where it's long been the dominant crab species in protected or somewhat protected sub-tidal and intertidal zones — areas such as bays, saltwater ponds and estuaries that are sheltered from the full force of the ocean's waves and currents. This range began to expand almost 200 years ago, when the species was first seen on the east coast of the United States and, not long afterwards, on the south coast of Australia. After a long period of gradual expansion, green crabs recently achieved Super Species status with major expansions on five continents.

With no way of making a transoceanic journey on their own, it's likely that green crabs have expanded their range so dramatically with human help. It's assumed that they've hitched rides aboard ships, as larvae suspended in ballast water or perhaps initially even as juveniles, clinging

to the mats of algae and barnacles that matted early ships' hulls. More recent shipping trends may be providing additional opportunities. For example, juvenile crabs are believed to get collected accidentally along with the seaweed used for packing and shipping live lobsters around the world. And crabs may also be moving between continents as part of the expanded global trade in products such as oysters and bait.

The bigger question is what makes this species so successful. Certainly green crabs have many of the attributes that seem to be prerequisites for successful invasion, including the ability to rapidly produce large numbers of offspring (one female green crab can lay up to 185,000 eggs per clutch and can produce clutches twice a year) and the biological tools needed to survive the environmental differences from one region to the next. These tools include an unusual tolerance for wide ranges of water temperature and salinity; a widely varied diet that includes mainly bivalves such as oysters, clams and mussels but also other crabs, seaweed, aquatic worms, insects and small fish; and an aptitude for learning new foraging techniques. Green crabs in their nonnative range also appear to be free of a particularly nasty parasitic barnacle — one that essentially converts the crab's innards into a barnacle-making factory, rendering the host sterile in the process — that plagues the species back home in Europe. Finally, the fact that green crabs are not particularly large, generally growing to no more than 3.5 inches (9 cm) across the carapace, means they're not being held in check by the deadly hand of human commercial exploitation.

While all these factors likely play a role, the ultimate reason behind the remarkable success of the green crab may be the fact that, in a changing world, they're simply better at being crabs than many of the native species against which they're competing. Recently scientists at Rutgers University in New Jersey set up a series of experiments to test this idea. In what amounted to a crustacean version of *American Gladiator*, the scientists collected green crabs, native blue crabs and members of a second alien

◀ Green crabs are able to outcompete other types of crabs at finding food, which is one of the reasons the species is so successful. Their success has caused heavy declines in populations of mussels and soft-shell clams — two of the green crab's favorite foods.

Female green crabs can lay hundreds of thousands of eggs in a single breeding season. Eggs like these may have initially been transported across seas in the ballast water of ships, explaining why green crabs are now found on five continents.

species, Asian shore crabs, to see how the three species fared in various head-to-head competitions over food. The results showed that the two invasive species were markedly stronger and meaner than native blue crabs of similar size.

When it came to procuring food, the green crabs were like heat-seeking missiles in their ability to outcompete the other two species. When individuals of each species were placed alone in a tank with a mussel, only half the shore crabs and 60 percent of the blue crabs were successful in locating the meal. By comparison, 95 percent of the green crabs passed the test. Even more impressive was the speed with which these crabs were able to zero in on their food. While it took successful blue crab and shore crab foragers an average of around nine minutes, green crabs needed barely more than 60 seconds. In direct competition, the larger-clawed Asian shore crabs were quicker to pick fights. Despite this, the green crabs used a strategy that was ultimately more successful. When two species were placed in the same tank with some food, green crabs challenged by aggressive Asian shore crabs would merely turn their backs and focus on eating. Even under such stressful conditions, virtually every green crab came away with a full belly. Their behavior could be explained by their physical characteristics: green crabs are more robust, almost tank-like, compared to the other two species.

Green crabs have been blamed for the collapse of a soft-shell clam industry off the east coast of the United States during the 1950s, and seafood harvesters around the world are concerned over the impact this species may be having on stocks of other shellfish and commercially important crab species. The challenge is how to handle the situation. To protect scallop populations off the coast of Massachusetts, local authorities in the mid-1990s put a bounty on green crabs. The result was the composting of more than 10 tons (9 tonnes) of the unwanted invaders. But as the spread of the species shows, nothing has been able to stop what appears to be an inevitable advance. Desperate researchers are now examining the wisdom of unleashing the green crab's native parasitic barnacle, and some food scientists are even trying to identify economically feasible ways of preparing green crabs for human consumption.

Of course, maybe in the end only nature holds the solution. One recent study reports that invasive Asian shore crabs may now be giving the green crabs a run for their money in certain habitats along the east coast of the United States. What's more, there is evidence that heavy predation of mussels by both these crab species may be driving the shellfish to evolve thicker shells, a finding that would add a dramatic new twist to our understanding of how Super Species such as the green crab are changing the world.

Part Two

EQUILIBRIUM LOST

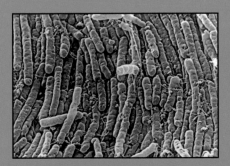

ECOLOGISTS, BEING SCIENTISTS, are driven by a search for laws. They seek general truths that can be summed up by equations that can be used to explain why something has occurred and to predict when it might happen again in the future. This can sometimes be problematic, because it inhibits them from dealing with something that may not be true all the time but is plainly so most of the time. One example of this is the ongoing effort to understand the concept of invasibility — the extent to which a given habitat is vulnerable to intrusions by invasive species. As long as ecologists have been concerned about invasive species — since 1958, when Charles Elton, widely regarded as the father of invasion biology, published his landmark book *The Ecology of Invasions by Animals and Plants* — they have noticed that nonnative species are good at gaining a foothold in some habitats and not so good in others. Much harder to figure out is why.

On the one hand, the answer seems almost too easy. As Elton first noted, there is a strong correlation between species invasions and habitats that have been disturbed in some way by human activity. Disturbance in this sense covers a range of impacts, the most obvious being permanent conversion of native habitats for other uses, such as clearing forests and draining wetlands for development and agriculture. Other human activities that disrupt or destroy local ecosystems include

- dam building (which alters the hydrology of watersheds)
- irrigation (which can deplete aquifers, alter river hydrology and increase the salinity of soil)
- fertilization (which leads to nutrient overload and oxygen depletion in marine environments)
- livestock grazing (which can denude grasslands, savannahs and shrublands of native vegetation)
- fire suppression and flood control (which remove natural disturbances that help shape the evolution of species and assembly of ecosystems)
- industrial pollution (which can alter air, water and soil chemistry and poison wildlife)
- burning fossil fuels (which may be altering global climate patterns)

- recreation (which can lead to a long list of problems, such as damage to coral reefs by scuba divers and crushing of sensitive vegetation by off-road vehicles)
- overharvesting or general persecution of wildlife (a direct hit on the integrity of native ecosystems, often with cascading side effects)

The link between invasions and disturbance is even easier to see today than it was in Elton's time. Run down the list of the world's most proficient and widespread invasive species, and chances are the habitats they've overrun have been greatly altered by some form of human impact. At the same time, undisturbed areas are rarely affected. In a study released in 2004 by WWF International, conservation managers from 34 countries were asked to identify threats to biodiversity in their respective preserves. Poaching, illegal logging and human encroachment were the most frequently cited problems. Invasive species barely received a mention.

Somewhere along the way, the important link between habitat disruption and invasion has been pushed aside. This is partly because of the argument that remote wilderness areas are free of invaders simply because they are remote, far from where humans tend to accidentally (or intentionally) introduce nonnative species. Some scientists may have been distracted by trying to identify the laws that drive the process of invasibility. Elton himself proposed that biodiversity is the key; the more species that inhabit a given environment, he argued, the harder it will be to invade. His hypothesis explains why islands seem to be more vulnerable to invasion than continents, but there is still no consensus over its validity as a rule. Some experimental results support the theory and others do not.

Finally, the lack of attention paid to disturbance as a factor may have nothing to do with biology. On the one hand, if human-mediated alteration of the environment is accepted as the underlying cause of proliferation of invasive species, the result is a rather inconvenient truth: in order to prevent invasions and protect native species, we humans have to clean up our act. If, on the other hand, we view invasions as a source of environmental degradation and threats to native species, then a more practical course of action presents itself: we must wage war on the invasive species. The latter, of course, is what we're doing, often with a hefty arsenal of chemical weapons. Compounding the problem is the fact that nonnative invasions, declines of native species and habitat loss often occur in the same place at the same time, which makes it difficult, if not impossible, for scientists to distinguish between cause and effect.

The role of disturbance has not been completely forgotten, however. The idea has recently rebounded in popularity, thanks in part to ecologists who are designing experiments to test how much habitat alteration factors into invasion success. One example is the work of Andrew MacDougall and Roy Turkington, plant ecologists who have been studying the Garry oak meadows of southwestern British Columbia, a biologically rich habitat of Garry oak trees scattered among open fields of grasses and wildflowers. On Vancouver Island only 10 percent of this habitat remains, and roughly half of what's left is dominated by invasive species; 144 different exotic plants now represent up to 75 percent of local plant diversity, with two nonnative grasses — Kentucky bluegrass and orchard grass — predominating.

To test whether the scarcity of native plants is a result of direct competition from invasive species, the researchers established study plots from which they selectively removed the two dominant exotic grasses over a three-year period. The scientists reasoned that if native grasses were being out-competed, they would return in force once the invaders were out of the way. But that's not what happened. Over the course of the study the savannah understory underwent a complete conversion, from a habitat dominated by grasses to one in which the main plants were native wildflowers and other forbs.

The conclusion drawn by MacDougall and Turkington is that the main problem facing native plants in the Garry oak savannah is not invasive species but habitat alteration, in this case fire suppression. The native plants that came to dominate the study plots are predominantly annual late-summer bloomers — plants whose life cycles evolved to help them avoid summer fires and droughts. Once humans began preventing fires from occurring regularly, these plants lost their advantage. Perennial grasses, on the other hand, are better suited for this type of environment, and left undisturbed produce a thick layer of litter that prevents other plants from establishing a foothold.

When the two researchers seeded sections of their plots with native grasses, they found that these plants grew just as well as the supposedly superior invasive grasses. This suggests the invaders dominate not because they are better competitors, but because current environmental conditions favor the spread of their seeds. (Most of the native grasses simply drop their seeds, and it is thought that Roosevelt elk, now no longer present in the remaining fragments of Garry oak habitat, may have once played an important role in long-distance dispersal. As the habitat became fragmented, this may have put these species at a great disadvantage compared to the wind-dispersed grasses being planted by humans around homes, farms and pastures.) What's more, grasses are thought to have played a vital role in maintaining the Garry oak savannah habitat in areas where fire traditionally occurred less frequently. This means invasive grasses, by filling in for the now-disadvantaged native grasses, may be playing a crucial role in preventing the habitat from being overrun by shrubs like the highly invasive Scotch broom — an ecological step that in other areas has led to the end of the savannah habitat altogether.

Additional evidence that human activity is a driving force behind invasions comes from studies done on fire ants, South American natives that have spread throughout the world and have been blamed for the disappearance of native ants. Biologists Joshua King and Walter Tschinkel established study plots in the pine flatwoods of Apalachicola National Forest, a native wilderness area in northern Florida containing rare longleaf pine trees and a high degree of biodiversity. When they mowed or plowed patches of native habitat, the researchers produced two interesting results. Native ants declined in both abundance and diversity, even when fire ants weren't present, and the fire ants showed a clear preference for disturbed areas. They never moved into undisturbed habitat on their own, but the invaders were all over the disturbed areas like rats in a pantry. The invasion was so overwhelming that the ants were establishing colonies faster than they could be removed. "These results," King and Tschinkel concluded, "demonstrate that human activity is the primary force that drives fire ant invasions and suggests that disturbance, not interspecific competition, has the greatest impact on structuring these ant communities."

Some researchers now believe that invasion is largely about opportunity.

According to this theory, a new species may invade a native ecosystem in response to a sudden excess of resources such as nutrients, light or water. This suggests it is possible for invasions to occur in pristine habitats, particularly those subject to natural disturbances that can create such opportunities. It also explains why invasions happen far more frequently and with greater intensity and duration in habitats that have been disturbed by humans. By saturating rivers and lakes with nutrients, by factory farming, by engaging in monoculture (such as cornfields, oyster farms and pine plantations), by clearing vegetation and forests, by discharging pollutants and, most dramatically of all, by creating new habitats such as reservoirs and cities, humans are generating ecological opportunities on a grand scale.

What's more, our input of resources into an environment is often continuous. We plant new seeds in our fields every year; fresh loads of nutrients regularly wash into the watersheds; roadsides and lawns are mowed regularly. For species that benefit from these resource pipelines, such actions represent virtually inexhaustible opportunities. This would explain the explosive and persistent nature of the worst invasions, from algae blooms to insect outbreaks to epidemics of parasites and pathogenic microbes. What makes the situation even worse is that once they are underway, the outbreaks themselves become a form of ecological opportunity and a potential source of habitat disturbance. The result is a positive feedback loop in which invasions beget invasions.

This is not to suggest that invasions never occur in pristine habitats. Introduce foxes to Alaskan islands filled with ground-nesting seabirds that have never seen a predator, as Russian and American fur traders did, and the result is bound to be ecological disaster. Similarly, moose introduced to Newfoundland more than a century ago have thrived in the absence of wolves. Among the damage they've caused is the spread of exotic weeds — dandelions, buttercups, hawkweed, thistles, foxglove — into remote protected areas of Gros Morne National Park. But even those weeds were found only around moose trails and the sites of recent insect outbreaks or fires — in other words, places where natural disturbances had occurred. The point is not that invasive species necessarily depend on disturbance, but that disturbance ignites invasions, often on a massive scale.

Greater recognition of this concept could have important implications. Invasive species would be seen not as a cause of habitat degradation but as a symptom of the way humans treat the environment. And this would make it illogical to blame invasive species for the loss of native biodiversity — like blaming the collapse of a rotting building on a strong wind. Even more important, it would force us to rethink how we deal with problems caused by invasive species. For example, drowning weeds in poison ignores the disturbance factors that may have precipitated the invasion in the first place, and it could make things worse by subjecting the environment to more abuse. In many cases, a better approach — for both revival of native species and control of invasions — may be better stewardship of the environment.

EASTERN GRAY SQUIRREL

Sciurus carolinensis

T**ALK ABOUT AMBIGUOUS** attitudes toward nature. To some people eastern gray squirrels represent one of the modern world's most endearing connections with the natural world. They're cute and cuddly; they seem perfectly at home in even the densest human settlements; and their backyard antics are a daily reminder that humans aren't the only species that knows how to solve a problem. To others, however, they're simply vermin. They eat flowers and dig up bulbs, nest in the attic and gnaw on electrical wires, raid bird feeders and chew the bark of treasured trees. In England in particular, the phenomenal success of this species as a nonnative invader has turned a nation of nature lovers into a horde of bloodthirsty killers.

Perhaps the mixed feelings stem from our recognition of how many characteristics we share with these otherwise dissimilar creatures. Gray squirrels are tree squirrels — a family of rodents that are the mammals best suited to exploit the resources of forest canopies in the temperate zone. In this regard, squirrels are to the vast forests of North America and Eurasia what primates are to the jungles of the tropics. Although different tree squirrel species can be found in a variety of forest types, each seems to thrive best under specific conditions. The small European red squirrel, for

◀ Clever and bold, gray squirrels have had no trouble adapting to living in close proximity to humans. In fact, being our neighbors has helped the species thrive in many ways; for example, we inadvertently provide the creatures with food on a regular basis.

example, spends almost all its time in the treetops and thrives best in dense pine forests. By comparison, the eastern gray squirrel spends more time on the ground and thrives in forests with lots of ground cover and large hardwood trees. They're experts at scavenging and storing fallen nuts and, perhaps because of this ability, they do well in tailored landscapes peppered with large, resource-rich trees such as oak, hickory, walnut, pecan, beech, hazel, maple and chestnut. Gray squirrels, in other words, like the same kind of forests we do.

Before they emerged as an invader, gray squirrels were already doing extremely well as a species. Their original range extended throughout the mixed deciduous forests of eastern North America, from the Gulf of Mexico to the southern parts of Ontario and Quebec. The dense untouched forests were so vast it was said that a squirrel could travel from the Ohio River to the Great Lakes without ever once having to touch the ground. Gray squirrel sightings were extremely common among the early European settlers.

As human populations grew in size and expanded their territory, the result was an increasing impact on the temperate biospheres, particularly forests. Initially trees were cleared for cities and farms, and later the need for timber and other wood products led to clearing or thinning of forests even in remote areas. Tree farmers, meanwhile, replaced native trees with those better suited for human needs. In many places the dense forests with their endless tracts of continuous canopy disappeared. What remained were both fragmented and often different from the native forests in size, shape, age, type and spacing of individual trees. Many native forest animals, including tree squirrels, were thus forced to deal with a forest infrastructure that was different from the one in which they had evolved.

The tree squirrel distribution map is now very different from what it was prior to colonial times, and two main factors have driven the transformation. The first was the straightforward decline of some species as their preferred habitat was eliminated. European red squirrels, for instance, are thought to have been common in Portugal from the Middle Ages until the end of the 16th century. As that nation became a seafaring superpower, however, the need for timber for shipbuilding and the clearing of land led to the species being driven to extinction locally. The second trend was a reawakening of the Western world's appreciation for nature, which has been traced to the rise of wealthy English and European elites with a penchant for large manors and stately gardens. One offshoot of this, made possible by the Europeans themselves invading other continents, was a widespread transfer of plants and animals as the landowners sought to enhance their new surroundings. Creatures that were seen as charming, unthreatening and hardy enough to survive were perfect candidates for satisfying such desires. On these counts, few species represented a more perfect match than the eastern gray squirrel.

During the 19th century members of the English upper class began importing gray squirrels from America. These were initially kept as pets, but they frequently escaped or were released outdoors. On dozens of occasions, gray squirrels were brought over a few pairs at a time from the United States or Canada and deliberately let loose in suburban parks or forests, with the intention of establishing new populations. This took place first across England and later in Scotland, Ireland and at least four places in the north of Italy. While not every introduction succeeded, many did. In England the species underwent a major population boom between 1930 and 1945, and it can now be found throughout most of the country.

In Ireland the first introduction is thought to have taken place in 1913, at a wedding reception on the grounds of Castle Forbes in Longford, involving 12 squirrels brought over from England as a gift. After surprising the new bride, the squirrels are said to have bolted for the nearest trees at their first chance. By 1923 the descendants of these squirrels had claimed territory that extended 10 miles (16 km) from the initial introduction site. In the following decades these animals (and possibly the descendants of at least one additional release) spread to occupy most of the country east of the River Shannon. Having recently been spotted west of this barrier, gray squirrels now seem poised to claim most, if not all, of Ireland

The gray squirrel's entry into Europe was much more recent. The first recorded introduction came after the Second World War, when two pairs were imported from Washington, D.C., by an Italian diplomat, who released them on the grounds of his villa south of Turin. The animals thrived, and by 2008 gray squirrels were occupying roughly 350 square miles (900 sq km) of northern Italy. Initially they were restricted to the suburbs, contained by extensive expanses of cultivated land with only small, scattered patches of trees. However, they've recently moved beyond this confinement with increasing speed, taking advantage of more continuous broad-leaved forests in the hilly areas of eastern Piedmont and the Po River valley. The invaders are now following a typical population expansion curve, and most experts agree they've reached the point where the population is ready to explode. Gray squirrels are expected to enter France and Switzerland within the next couple of decades, if not sooner. Much of Europe, it seems, lies within range.

On other fronts, introduced gray squirrels have had mixed results. Animals set loose in South Australia by English settlers in the late 1800s resulted in a population that survived for a century before slowly dying off in the 1970s. On the other hand, gray squirrels introduced near Cape Town, South Africa, in the early 1900s have spread throughout urban, agricultural and reforested areas in a 2,700-square-mile (7,000 sq km) swath of Africa's southern cape. The species also appears to be well on its way to establishing major permanence in the western parts of North America. In the early 1900s gray squirrels were introduced to Vancouver's Stanley Park, and in recent decades they have spread throughout the lower Fraser Valley. On Vancouver Island, two female gray squirrels and a male that escaped from a game farm in 1966 are believed to be the founders of a population that grew slowly before expanding rapidly in the mid-1980s. Now gray squirrels can be seen throughout most of the populated southeastern portion of the island and as far north as Nanaimo. Other thriving gray squirrel populations can be found in and around Calgary, Seattle, Portland and San Francisco. As in England, many western North Americans now view these expanding populations with increasing unease.

Why has the species done so well? One reason appears to be linked with reforestation. Having extensively cleared native forests during the first centuries after colonization, Europeans and their descendants have in recent decades become increasingly surrounded by trees. Old hardwoods in city parks have matured. Other large trees now fill suburbs, commercial zones and the sides of roadways that were razed for development during the postwar expansion of the 1950s and 1960s. This has been further bolstered by a growing appreciation for environmental health that has become more widespread in Western culture since the 1980s. What has no doubt made this doubly advantageous for animals like squirrels is the fact that, while the trees have returned, the predators have not, at least not in large numbers.

Of course, at the root of the gray squirrel's success are traits that enable this species to thrive in modern habitats to a much greater degree than any other squirrel species and most other mammals. One is an ability to forage and store food. It's true that such skills are common among rodents, but gray squirrels seem especially adept. As tree squirrels that are also at home on the ground, they're able to collect large quantities of fallen nuts and to bury them over a wide area. Studies have revealed that they're also capable of pilfering food that's been stored by other animals in their territory. Such provisioning skills are thought to be vital for survival in the temperate zone, where having a ready food supply in the winter can be crucial to ensuring that adults are healthy enough to meet the energy demands of rearing offspring in the spring. Indeed, gray squirrels are among the squirrel species capable of producing two litters a year — an admirable feat given the shorter window presented by their seasonal climate.

Gray squirrels also possess a number of traits that provide countless hours of amusement for homeowners and park visitors. They're bold, which enables them not only to run the daily gauntlet of cars, cats and pedestrians but also to even climb up into laps or stick their nose into a pocket for a hidden peanut. Their amazing agility and intelligence, meanwhile, are well-known to anyone who has ever matched wits with them in an effort to prevent raids on bird feeders. All of this is accentuated by the fact that gray squirrels are also able to eat a wide range of foods. Nuts and seeds are their mainstays, and scientists have found that gray squirrels make specialized enzymes for digesting substances in acorns that other squirrels cannot utilize. When nuts are in short supply, gray squirrels derive sustenance from eating flowers, buds, fruit and mushrooms, as well as eggs and insects. Thanks to their iron guts, gray squirrels are also at home in a world of french fries, bagels, popcorn and many other scraps and handouts associated with humans.

Not surprisingly, a large urban park with lots of leftovers, garbage cans, oak trees and plenty of noise to scare away timid competitors is for a gray squirrel nothing short of utopia. One study done in 1980 found that gray squirrel densities in Washington D.C.'s Lafayette Park amounted to roughly 20 animals per acre (50 per hectare), more than ten times what one would normally find in the wild. Other studies have shown that as one moves from rural forests to suburban parks to the inner city, gray squirrel populations get denser and denser.

One might think their success would earn gray squirrels kudos as true survivors, but that has not been the case — quite the opposite, in fact. The more success these animals have had in their new territories, the more that human attitudes have shifted from adoration to loathing. In England, wildlife authorities, private landowners, foresters and friends of native species have taken part in a decades-long rampage in which tens of thousands of gray squirrels have been shot, poisoned and bludgeoned to death with all manner of heavy objects. More recently there have been attempts at developing a more humane plan: to sterilize gray squirrels with baits containing agents that interfere with fertilization. But the end goal remains the same — total extermination.

Scottish conservationists got into the act in the spring of 2009, when they launched a three-year cull expected to cost about $2.2 million. Naturalists and landowners are joining forces in an effort that's expected to wipe out tens of thousands of gray squirrels out of an estimated total population between 200,000 and 300,000. Throughout the British Isles the challenge is being met with a fervor at times tinged with glee. Typical responses to a recent article published online by *The Daily Telegraph* in London included this one: "This is a no-brainer. Persecute them all to extinction." Another recent commentator said: "Personally I think that all gray squirrels should be shot on sight, and if it is safe to do so anyone with a spade, air rifle or shot gun should show them no mercy."

This dedicated opposition to gray squirrels results from a campaign to educate people about the implications of the invasion. Advocates of eradication argue aggressively that the squirrels, by preying on eggs, threaten native songbirds. By chewing on bark, the invaders have also been declared guilty of destroying forests and inflicting huge economic losses on foresters. But the primary fear is that gray squirrels threaten native species, particularly the European red squirrel. In this regard gray squirrels are seen as aggressive animals that quickly displace the smaller red squirrels by chasing them away from suitable habitat and killing their young. They're also known to spread a virus that kills red squirrels but spares their own kind. All this has led to the widespread belief that if the gray squirrels are not destroyed, the reds will be driven to extinction. The emotional resonance of this narrative is particularly strong because red squirrels are English cultural icons, immortalized by Beatrix Potter as the beloved character Squirrel Nutkin. (Timmy Tiptoes, Potter's gray squirrel character, never achieved the same popularity.)

The accusations leveled at the gray squirrel are often highly exaggerated. Gray squirrels do chew bark, and this can result in unsightly scars, reduction in the value of the tree as a source of timber and sometimes even tree death (up to 5 percent of a woodland in the worst cases). But there is no evidence that gray squirrels threaten native forests. Much of the damage they do inflict, meanwhile, may be a result of the practices of British tree farmers. Evidence for this comes from the observation that gray squirrels do not cause extensive tree damage in their native range. In Britain they have been found to hit tree farms harder than naturally regenerating forests; there is evidence that they're attracted to stands where thinning and regular fertilizing have spurred rapid growth and tastier inner bark. The charge that gray squirrels are driving songbirds to extinction is also driven more by emotion than fact. Gray squirrels will eat eggs and young birds when given the chance, but there's no evidence that this is a major threat. Songbird populations were in decline before the extensive spread of gray squirrels in Britain, and they've also declined in parts of Europe where there are no gray squirrels. These trends are thought to be due to loss of suitable habitat. Further, in early 2010 the *Journal of Ornithology* published a major analysis of 38 bird species across England, which failed to find any evidence linking gray squirrel success to bird declines.

Ironically, red squirrels also chew bark and eat eggs when given the chance. In fact, by the late 1800s, red squirrels, which had declined in

The destruction of pine forests such as this leaves native species like red squirrels without a home. Forced to live in different habitats, red squirrels have more trouble finding food and are easily outmatched by hardier gray squirrels.

number following deforestation in the previous century, underwent a resurgence. They eventually came to be regarded as pests, bounties were established, and the squirrels were widely hunted. Groups such as the Highland Squirrel Club, formed in Scotland in 1903, came into being for the sole purpose of ridding local pine forests of this menace. By 1917 members of this organization alone had killed more than 60,000 red squirrels.

As for displacing native squirrels, gray squirrels are indeed guilty. During the last half of the 20th century a familiar pattern occurred across much of England and Wales: gray squirrels moved into an area and a decade or two later the red squirrels were gone. But this relationship is now known to be more complex than first thought. For one thing, gray squirrels don't pose any direct threat to reds. They don't chase them; they don't eat their young. However, they do monopolize resources under certain types of forest conditions. This brings us back to the fact that different squirrels are adapted to subtle differences in forest habitat. In forests with lots of oak trees, for example, gray squirrels feast on acorns and make it through to spring in good health. Red squirrels, by comparison, have a tougher time in such forests. If there are no gray squirrels present they're able to eke out a living and survive at low densities. When gray squirrels move in, however, the struggle becomes too much for the red squirrels: energy spent on finding food and shelter doesn't produce the energy required to reproduce successfully. As a result, their population dies out.

But the parts of this equation can change along with the type of forest. In dense pine forests, the smaller red squirrels are well adapted for scampering out onto thin branches to collect nuts that are inaccessible to the heavier gray squirrels. This would suggest that smaller squirrels are better suited for this particular habitat, and it would also explain why the pine forests in the north of England and in Scotland appear to be one battleground where the red squirrels are still holding their own. Indeed, red squirrel numbers in some parts of Scotland have even been expanding in recent years.

Some scientists and others argue that trying to eradicate the gray squirrel from Britain, Europe and western North America is ultimately futile, and thus a waste of time and effort and damaging to society's moral fiber. According to this line of reasoning, the increasing success of gray squirrels is ultimately our fault: we've created a playing field upon which these highly adaptable rodents rule supreme. This means that the best long-term approach to saving species such as red squirrels lies in efforts to preserve native forests and changes in the way we husband the land — not in shotguns and hammers.

As for gray squirrels, there seems to be no easy answer for the ambiguity of our love/hate relationship. We'll always have city parks and backyard gardens, which means that unless we're willing to make a Herculean effort, we'll always have gray squirrels. And in the absence of predators, we're also likely to have too many of them for people's liking. In Britain there's a new drive to encourage humans to eat squirrel meat, a common practice in the southern United States. Whether the trend will take off remains unclear. Either way, this phenomenally successful species seems destined to remain a part of nature that leaves us torn over what it is — a cute and cuddly pest.

KILLER ALGAE

Caulerpa taxifolia

LET'S FACE IT: algae don't have a lot of charisma. These primitive organisms are best known for their contributions to more elaborate life-forms: as symbiotic partners in lichens and coral reefs, or as the evolutionary precursors to true land plants. On they're own they're simple organisms that lack much in the way of specialized form or function — components of the biosphere that are easily forgettable. One notable exception, however, is *Caulerpa taxifolia*, a form of seaweed also known as "killer algae."

In the 1960s, staff at the Wilhelma Zoo in Stuttgart, Germany, began using *C. taxifolia* to decorate its saltwater aquariums. The seaweed, which is bright green with delicate-looking fronds, was not only beautiful to look at but also easy to work with. For one thing, it was hardy. A tropical alga normally found only in warm waters, this particular strain showed an unusual tolerance for cold water. Its good looks and versatility soon made *C. taxifolia* something of a celebrity among aquarists. The alga became part of the trade in fish-tank products and was soon being shipped around the world.

But that's only the beginning of the story. In the summer of 1984 a small patch of *C. taxifolia* was seen growing in the Mediterranean Sea outside the Oceanographic Museum in Monaco, a prestigious institution formerly run by the famous French oceanographer Jacques-Yves Cousteau. To this day no one knows how it got there, but the aquarium was using *C. taxifolia* at the time, which (along with later genetic tests) strongly suggests that it was either accidentally or intentionally released. At any rate, it somehow escaped.

At first its presence remained contained and mostly a secret. But in 1988 the invading alga came to the attention of a marine biologist at the University of Nice, Alexandre Meinesz, who is an expert on *Caulerpa*. Meinesz was immediately amazed by how the seaweed had managed to survive at least five years in what for a tropical species should have been fatally chilly winters. His concerns about its potential threat to native ecosystems, however, fell on deaf ears. The invading alga, meanwhile, began to spread with a speed that was eventually impossible to ignore. By 2000 it was estimated to be blanketing 30,000 acres (13,000 ha) of seabed off the shores of six Mediterranean countries. Around the same time, divers began finding the same strain of *C. taxifolia* off the coasts of California and southeastern Australia. By that point the media had latched on to the story, and word began to circulate that no temperate marine zone on the planet was safe.

There is some irony surrounding this alga's success. Although widely distributed throughout its natural habitat — it can be found in tropical oceans around the globe — *C. taxifolia* usually forms only thin, scattered patches, making it hard for divers to find it. In the Mediterranean, however, the story is quite different. Researchers have seen the algae growing in vast, dense meadows blanketing a variety of seafloor habitats from sheltered bays to underwater cliffs to the storm-pounded seafloor near headlands. It's also been found at a surprising range of depths, from water that is only waist-deep to just over 300 feet (100 m) into the murky depths — places where most photosynthetic life-forms would find it impossible to eke out living.

As one of 86 members of the genus *Caulerpa*, *C. taxifolia* belongs to a clan that is distinct in several ways from most other algae. Consider their architecture. The main part of the *Caulerpa* body is a stolon, which is a horizontal stem-like structure that grows along the seafloor like a snaking vine. Shooting up from the stolon is a series of fronds bearing leaf-like blades

These scientists are covering a patch of *Caulerpa taxifolia* with a tarp ▶ and will then inject the Super Species with liquid chlorine. This seems to be an effective — albeit expensive — method for controlling the algae.

for gathering light; growing downward is a network of root-like rhizoids. In other large algae such as kelp, these latter features are used for anchoring the algal body to the substrate, but among the *Caulerpa* they perform double duty. In addition to their attachment function, they also evolved a means to extract nutrients from sediments, just like the true roots of plants. What's more, *Caulerpa* rhizoids harbor symbiotic bacteria that are thought to aid in transformation of elemental nitrogen into more usable forms. This is a far from trivial ability. Nitrogen is an essential element required by all plants and animals and is the major component of Earth's atmosphere. However, the bulk of this nitrogen is unusable by most life-forms unless it is first converted to absorbable nitrogen-based compounds, a process that's usually carried out by bacteria. Thus, while most large-bodied algae must be content to absorb nutrients from the water, members of the *Caulerpa* clan are not only able to exploit sediment nutrients but also thrive under conditions that many other life-forms would find hostile.

Caulerpa are distinctive in other ways too. They're highly toxic, producing a cocktail of poisons that is thought to be an effective deterrent to would-be predators. And they have a very peculiar makeup at the microscopic level: each algal body, including stolons, fronds and rhizoids, consists of just one cell — one continuous, delicate tube of organelle-filled cytoplasm. With some outcrops of *C. taxifolia* consisting of hundreds of fronds attached to a stolon that can be up to 9 feet (2.8 m) long, these unusual creatures stand among the largest single cells in the biosphere. Besides earning them points as bona fide freaks of nature, this unusual trait appears to contribute to the algae's ability to cheat the normal laws of growth, and even death. For one thing, each giant *Caulerpa* cell can grow in all directions — down into the substrate, horizontally along the seafloor and up toward the surface — and it can do so with impressive speed. In the Mediterranean, *C. taxifolia* has been found to lengthen by more than 3/4 inch (2 cm) each day during its peak summer growing season. Its life-cycle pattern, meanwhile, is what experts call pseudoperennial: as one end of the stolon dies, the other continues to grow. Unharmed, this seaweed could theoretically live forever.

Perhaps not surprisingly, several members of the *Caulerpa* genus have emerged as invasive species. Some of the more noteworthy examples include *C. racemosa*, which has also invaded the Mediterranean, spreading from the coast of Libya and into the waters of 10 additional countries since 1991; *C. brachypus,* which was first seen near Florida in 2001 or possibly earlier; and *C. scalpelliformis*, which appeared in Botany Bay, near Sydney, Australia, in 1994. But none of these invasions have matched the success of *C. taxifolia.*

Why? One reason may be toxicity. Tests done on the invasive strain of *C. taxifolia* have revealed exceedingly high levels of poison compounds, even for *Caulerpa.* Caulerpenyne, the principal toxin carried by these algae, has been found to represent up to 13 percent of *C. taxifolia*'s frond weight during the summer. Further studies have found that much of this poison is stored in the form of unassembled molecular building blocks. When a fish or other grazer takes a bite, these chemical pieces are released and assembled in the surrounding water, turning it into a poisonous soup. Aquarium operators have observed that tubeworms — worm-like animals that live inside sleeve-shaped shells — will crawl out of their shells and die when put in the same tank as damaged *C. taxifolia.* Lab experiments have shown that sea urchins, creatures that normally thrive on seaweed, will nibble on their own waste or even bits of plastic, eventually starving to death, rather than touch the alga (at least during summer, when toxin production is at its peak). There have also been reports of humans experiencing hallucinations, vertigo and loss of memory after eating fish thought to have grazed on *C. taxifolia.* While some tropical fish and other organisms are known to have evolved immunity to this weapon, most Mediterranean species are thought to simply avoid this seaweed.

This toxicity has earned *C. taxifolia* its reputation as the killer algae, but it likely isn't the only secret to its success. The species is also extremely adept at reproducing. Many algae replicate both sexually (by releasing a reproductive cell that combines with a similar cell from another individual) and asexually (when a fragment breaks off and sprouts in a new location). Oddly, scientists have failed to find any evidence that the invasive *C. taxifolia* strain has spread sexually, which suggests that the Mediterranean has been overrun by the offspring of a single male. What has been observed is that the alga appears to be expert at spreading asexually. In lab studies, new bodies of *C. taxifolia* sprouted from severed fragments barely more than a 1/3 inch (10 mm) in length, a quarter of the minimum size required by other *Caulerpa* species tested.

Furthermore, *C. taxifolia* appears to be difficult to kill. When a piece of the seaweed gets wrapped around a boat anchor, sucked up with ballast water or even washed up on a beach, it is still able to return to the seafloor and resume proliferation, despite the physical abuse. One study found that severed fragments can survive out of water for up to 10 days. Such findings suggest that the species can be readily exported to new locations by human activity.

Yet another idea regarding the alga's success — one that has received a lot of attention since its spread first came to light — centers around the possibility that it acquired additional fitness traits during nearly two decades of selective breeding in captivity. Strong initial support for this theory arose from the alga's unusual tolerance for cold waters. Most native populations of *C. taxifolia* are found in parts of the world where sea surface temperatures rarely dip below 68°F (20°C). In the northern Adriatic Sea, the coldest region invaded by the aquarium strain thus far, winter sea surface temperatures drop to below 34°F (10°C). Studies have shown that the invasive strain can survive for three months under such frigid conditions. Recently, however, genetic studies have suggested that the invasive strain of *C. taxifolia* may have derived from a native population that lives in subtropical waters near Queensland, Australia, where winter water temperatures are often similar to those in the northern Mediterranean. This suggests that nature should get the credit for having selected such a cold-tolerant strain. Whether selective breeding is responsible for any fitness traits linked with *C. taxifolia*'s success — whether it's a rampaging mutant, in other words — remains to be determined.

While some scientists continue to search for the secrets of its success, others have been busy looking for ways to control its spread. This has not been easy, particularly in the Mediterranean, where control efforts often did not begin until the alga had become well-established. Government authorities have considered eradication proposals that include blowtorches, bell jars pumped full of boiling water, infusions of bleach, underwater cement, a shower of copper ions, and dry ice. One community on the coast of France tried to protect its harbor by covering a newly discovered patch of *C. taxifolia* with black plastic sheeting held down by metal wire netting. Six months later, the alga was thriving, having taken root in the new layer of sediment that had settled atop the plastic.

Since the algae spread in a way that is patchy rather than continuous, it has emerged that the best way to deal with the problem is to hit the seaweed when new patches are in the early stages of colonization. This strategy was successful in some European locales, where small squadrons of scuba divers laboriously ripped out recently established patches by hand. American authorities were also quick to act when *C. taxifolia* was first spotted in U.S. waters, first in Agua Hedionda Lagoon near San Diego, in June 2000, and then a month later in Huntington Harbour, near Los Angeles. A well-coordinated effort between government agencies and community organizations resulted in a $6-million eradication campaign in which the algae were covered with plastic tarps held down by sandbags and simultaneously doused with bleach and other biocides. Success came almost immediately, and as of 2009 there were no signs of *C. taxifolia* in the continental U.S. As for the well-established patches in the Mediterranean, one solution may be a particularly voracious sea slug that doesn't seem to mind the alga's nasty chemicals. Divers around the world are now being warned to keep an eye out for *C. taxifolia*, and other measures are being taken to control its spread. In the United States, for example, laws have been passed making it illegal for individuals to own or sell nine different *Caulerpa* species, including *C. taxifolia*.

Despite all this, one important chapter in the tale of the killer algae still remains largely unwritten: the one on the ongoing debate over whether *C. taxifolia* is anywhere near as big a problem as is widely believed. Right from the start, Meinesz and others have argued that the spread represents a major threat not only to native species — including commercially valuable fish and shellfish populations — but also to the very integrity of native marine ecosystems. This belief is supported by French studies in the early 1990s that illustrate the total dominance *C. taxifolia* can exert over its environment: when it invades an area, its increasingly dense fronds hog

Although most species avoid eating this highly poisonous alga, the violet aeolid sea slug (*Flabellina affinis*) is immune to the its toxins. The sea slug presents a potential form of biological control against the algae, but unleashing another alien species may do more harm than good.

most of the light; as a result, plant and algal diversity in the bottom zone is greatly reduced. One study found that *C. taxifolia* meadows contain fewer marine invertebrates such as mollusks, insects and worms.

There is also evidence that the seaweed harms native sea grasses, the marine vegetation that normally dominates the near-shore environment in the Mediterranean. Researchers have noticed, for example, that once the invasive algae move in, sea-grass leaves grow less vigorously and a substantial number of shoots eventually wither and die. Although the full extent of this damage in the wild isn't clear, any threat to sea grass is widely seen as a serious threat. This native vegetation is the cornerstone of seabed ecosystems. It stabilizes fine sediments that would otherwise get flushed away by currents, protects beaches from erosion, and supports the food web. As the habitat's main structure for thousands of years, it's the environment to which local native species are best adapted. It's been noted that native octopi and certain fishes blend in with the muted colors of the sea grass but stand out starkly against the bright green of *C. taxifolia*.

Finally, there are additional concerns that the highly toxic nature of *C. taxifolia* will effectively scare away fish, sea urchins and other herbivores. An informal survey of Italian fisherman found that 77 percent reported catching fewer fish in the years after *C. taxifolia* had invaded their traditional harvest zones.

But not everyone is sold on the idea that *C. taxifolia* is as dangerous as widely believed. Administrators at the Monaco aquarium — perhaps eager to escape the blame for what others have described as an ecological disaster — have long maintained that *C. taxifolia* is primarily drawn to polluted or otherwise disturbed habitats. Its rapid spread, they argue, isn't due so much to the superior fitness of the algae as it is to the disturbed nature of the marine environment. This includes dumping of sewage, fertilizer runoff and other forms of pollution, as well as overharvesting of marine resources, the introduction of oyster farms and other aquaculture operations, and the physical damage caused by boat anchoring and certain fishing practices. According to this line of reasoning, human disturbances weaken native ecosystems while at the same time providing an environment that's conducive to the growth and spread of invasive algae. Proponents go so far as to argue that *C. taxifolia* may actually help repair polluted habitats by removing the excess nutrients introduced into the environment by humans and by creating habitat where native vegetation can no longer grow.

Several compelling observations support this idea. Many algae thrive at the expense of aquatic plants in marine habitats that have been contaminated by high levels of nitrogen and other nutrients found in sewage and fertilizer. Indeed, pollution was hypothesized to be the culprit in an earlier spread of *Caulerpa* algae that occurred on the French Riviera in the 1920s. More recently, researchers monitoring the spread of *Caulerpa* in places such as Florida and the Bahamas have also noted a correlation between the alga's success and the degree to which local waters have been enriched by nitrogen from human sources.

Currently in the Mediterranean, researchers see a similar situation. For example, native sea-grass beds were known to be in decline in some areas for at least 30 years before the arrival of *C. taxifolia*. Many of the worst invasion sites seem to be in areas where the sea grass is dying or no longer grows. A recent study involving detailed aerial surveys of six sites along the French coast offered further support. The results, published in 2003, led the authors to conclude that the total area covered by the algae, at least within the survey zone, was much less than what had been previously estimated. The researchers also noted that some of the densest areas of algae growth were in harbors around sewage outlets, storm-water drains and places where the seafloor had been torn up by anchors, ground-fishing equipment or other human activities. By contrast, the comparatively pristine waters around the island of Porquerolles were almost algae-free. The one significant patch of *C. taxifolia* there was found growing in an area where the sea grass had been destroyed and contaminated by underground weapons tests conducted by the French navy. In a recent review of the situation, Italian researchers concluded that "polluted waters and general human impact on the coastal zone played a leading role in lowering competitive ability of native seagrasses, permitting the invader to dominate in the environment."

Meinesz, for one, remains unmoved. In many areas he's seen *C. taxifolia* take hold at the edges of healthy sea-grass beds, and he argues that this is just part of a long-term process that could harm otherwise stable ecosystems. Unfortunately, the many difficulties associated with scientific study of complex ecological processes make it hard to say who's correct. How, for example, does one define what a disturbed habitat is in an area that's been the center of human civilization for thousands of years? What is clear is the importance of continuing to strive for answers. At stake is whether millions of dollars spent on eradication efforts represent money well spent or a complete waste of resources. Hopefully, time will tell.

FERAL PIG

Sus scrofa

How is it possible that every list of invasive species worth its salt invariably includes that lumbering beast known to scientists as *Sus scrofa,* and to the rest of us as the domestic pig? Here's an animal we've kept penned up on farms for thousands of years and transformed through selective breeding into a highly prized piece of walking meat that seems to lack any of the instincts that once helped it deal with such wilderness menaces as wolves and tigers. Look at it: it's fat, it can barely move, it's utterly defenseless and . . . did someone mention it's extremely tasty? Surely this floppy-eared grunter couldn't invade anything more challenging than a feeding trough.

But looks can be deceiving, even in the world of invasive species. It turns out the humble pig is one of the most destructive animals ever to set hoof on foreign turf. They eat and trample crops; they destroy the forest floor and transform vegetation across entire landscapes; they spread disease; they even eat native species. What's even more alarming is the degree to which the species is thriving. In recent decades feral pigs have rapidly moved into new territory in Europe, Asia, North America and Australia. At the same time established populations have increased dramatically, despite more and more concerted efforts aimed at keeping them under control. Almost everywhere you look, pigs are running wild.

The original *S. scrofa* is the wild boar, beast of legend and companion of the ancient Norse gods. It originally ranged from England to Japan and from Siberia to Indonesia, which suggests that humans have long had

contact with pigs over much of the globe. Indeed, recent DNA evidence suggests that, unlike most other domestic animals, humans have tamed wild boars multiple times in multiple places — including areas now part of Italy, India, Burma, western Indonesia and China — dating as far back as 9,000 years. But the relationship between man and pig went beyond domestication. As humans began traveling the world they took their trusty sidekicks almost everywhere they went. Some pigs were kept close at hand but others were released into newly discovered wilds.

The seafaring Polynesians were responsible for bringing nonnative wild pigs to just about every major island chain in the Pacific, including, about a thousand years ago, the islands of Hawaii. The Europeans later followed suit. Captain James Cook presented pigs to the Maoris on his first visit to New Zealand. English colonists aboard the First Fleet had 32 pigs with them when they sailed into Sydney's Botany Bay in 1788. Shortly afterward, feral pig populations were established in New South Wales, as well as outside the settlements established one after the other in coastal regions around Australia. In the Americas, Spanish conquistadors introduced pigs in 1539 to what is now the state of Florida, while English settlers later brought them to New England. The ensuing centuries saw repeated introductions and escapes of domestic pigs, as well as intentional releases of wild boars wherever humans desired a ready supply of meat. All of these pigs are thought to have interbred freely to produce the mixed gene pool that makes up today's feral pig population.

The release of wild pigs was often a prudent move for early human migrants. *S. scrofa* is an extraordinarily adaptable creature. It is capable of surviving in all kinds of habitats, including temperate hardwood forests, rain forests, marshlands, grasslands and even semi-arid rangelands, provided there is some water available. Part of this adaptability is due to its being able to withstand a range of climates — pigs are as at home in the cold boreal forests of northern Russia as they are in the steamy rain forests of Australia and the parched chaparral of south Texas. Just as important is

◀ It's hard to imagine these cuddly-looking piglets could pose any major threat. However, invasive wild boars wreak extreme damage on forests and grasslands around the world.

the fact they can eat almost anything. Like humans, pigs are consummate omnivores. They love fruit and nuts, but they'll also eat leaves, stems, shoots, roots, bulbs and mushrooms. And what many people don't realize is that they're also fond of meat and other animal protein. Given the chance, pigs will eat eggs, worms, snails and rotting carcasses. If the chance arises, they'll even eat live animals. On islands in the South Pacific, feral pigs are known to dig out and eat burrowing birds such as petrels and consume the eggs, nestlings and adults of ground-nesting birds. One historical account from Rarotonga in the Cook Islands describes pigs as an excellent form of rat control. It was because of this versatility and adaptability that early pioneers felt confident that pigs would not only survive when released in the wild, but thrive. You could call them a low-cost, low-risk insurance policy against starvation.

The status of *S. scrofa* over the centuries has not been well documented, nor has its range always been expanding. From the Middle Ages onward, for example, wild boars in many parts of their native range came under increasing pressure from hunters as human populations grew across Europe and Asia. This led to eradication of the species from certain regions, including all of Scandinavia and the British Isles. Its current revival appears to be taking place on several fronts. A combination of escapes, releases and expansion of surviving populations is thought to have contributed to a major resurgence of wild boars throughout Europe that has been going on for the past 50 years. Populations have been reestablished in England and Sweden, and growing numbers of wild boars have been reported throughout most of the rest of Europe. In Russia a massive expansion of wild boar populations has resulted in their taking over most of the country up to the Siberian border. While some Russian biologists think this expansion represents repopulation of the species' former range, others think wild boars now occupy more territory than ever before.

In Australia feral pig populations have gradually taken over more and more land. They've moved into Kakadu National Park, the renowned wilderness area of Crocodile Dundee fame. They've infiltrated the rain-forest region of northern Queensland. They can be found in almost every coastal area and in many inland areas where water is available, a territorial land-grab covering nearly 40 percent of the continent. That's impressive, given the fact that the country is mostly desert. But what's truly mind-boggling is the size of the populations: estimates suggest that the number of feral pigs in Australia ranges between 13 and 23 million, which means there's a chance there are now more wild pigs than humans.

North America appears to be headed for the same fate. Traditionally confined mainly to the southern states, feral pig populations are now expanding, rapidly invading new territory on several fronts, according to the Northern Feral Pig Project, a loose organization of wildlife officials from the northern U.S. and Canada that has been formed to document the spread. During the past five years sightings have been reported with increasing frequency from far northern states such as Michigan, Minnesota and North Dakota, as well as from parts of British Columbia, the Canadian Prairies and Quebec. Today wild pigs are well established in 21 U.S. states and have been seen in 13 others, with a total population recently estimated to be somewhere between two and six million — and rising. In Alberta, populations have become so large that provincial legislators passed a law in 2008 declaring them to be pests. The pigs also run wild in their traditional southern strongholds. In California they've invaded new areas, including the Santa Cruz Mountains. In the late 1990s the state began hiring professional trappers to reduce out-of-control pig populations in state parks, including some close to San Francisco.

What's going on? On the one hand, credit must go to the pig. Despite sometimes being portrayed as dim witted, *S. scrofa* is a highly intelligent creature. This is one reason why the animals frequently escape from captivity. It also partly explains why in many places they can capitalize on food resource opportunities presented by farms and, increasingly, residential areas — they're good at nighttime raids and getting around fences and other barriers. Pigs are also quite bold and, despite their reputation, aggressive. Such traits, along with their adaptability, have enabled feral pigs to thrive in a world increasingly dominated by humans.

It doesn't hurt that they also breed like rabbits. Female feral pigs in the wild can produce up to two litters a year, with as many as 13 piglets per litter (more than any other ungulate — indeed, more typical of small mammals like rabbits). The highest breeding rates occur when feral pigs carry the blood of domestic pigs, a fecundity likely reflecting centuries of artificial selection to create perfect piglet-making pork machines. Either way, the output is impressive. Australian wildlife officials crunching the numbers have come to realize that if a feral pig population were reduced by 70 percent — roughly equivalent to the easy part of reducing a population of animals in the wild — absence of further controls would see it return to full force in as little as two and a half years.

The second part of the equation — the other side of this increasingly unbalanced ecological balance sheet — is that feral pigs are running free in a world where there is very little to keep their numbers under control. In this regard, humans are directly to blame. This is partly because we no longer depend on hunting wild animals for food, at least not to the degree we did during earlier centuries, when wild boar numbers were reduced throughout Europe and kept under control in places like Hawaii. Just as important, during the past century we have erased a good majority of the large predators that would have kept populations of animals like pigs under control. Although small predators such as foxes and alligators still prey on piglets and the occasional adult, the decimation (and sometimes disappearance) of wolves and the big predatory cats throughout North America and Eurasia has turned these continents into mostly worry-free zones for free-range pigs and their offspring.

The Pacific islands, including New Zealand and even Australia, despite its dingoes, were pig playgrounds right from the start. In many other places, big predators are on their way out, largely because remaining wilderness preserves are too small and fragmented to support viable populations of large animals. Malaysia's 10-square-mile (2,590 ha) Pasoh Forest Reserve is a good example. Despite being a protected area, this lowland tropical forest preserve is no longer home to tigers or any other large predators, and today the wild pig population is out of control. According to one recent study, the *s. scrofa* population has increased to the point where it is now somewhere between ten and a hundred times larger than ever.

None of this would be a problem if not for the fact that out-of-control species inevitably wreak environmental havoc. In addition to being voracious eaters, pigs are also powerful animals, equipped with muscular necks, wedge-shaped heads and highly dexterous and capable snouts. These features explain why they're such fanatical diggers. Indeed, few creatures are more capable of meeting the energy demands required to upturn huge volumes of earth in pursuit of roots and bulbs. And in case you're thinking of Babe from the movies and wondering how such an adorable creature could do any environmental damage, think again. Think Hogzilla, a now legendary feral pig that was shot in Georgia in the summer of 2004. An examination of its remains led scientists to conclude that this animal, which had tusks nearly 18 inches (46 cm) long, measured about 8 feet (2.4 m) in length and weighed around 800 pounds (360 kg), although the owner of the land where the beast was shot still maintains that it was 12 feet (3.7 m) long and weighed 1,000 pounds (450 kg).

Quibbling over world records aside, the point is that feral pigs represent fleshy Rototillers capable of disrupting terrestrial habitats on a scale that's almost hard to comprehend. One study of grassland meadows in California found that feral pigs were capable of disturbing 7.4 percent of the ground surface over the course of a year, whereas native species — moles, pocket gophers, ground squirrels, skunks and ants — affected less than 1 percent of the soil combined. In Florida, researchers reported discovery of a power-line corridor in Savannas Preserve State Park that looked as if it had been carpet-bombed. The 2.2-acre (9,000 sq m) area, which at one point had been cleared of its original vegetation and replanted with easier-to-maintain grasses and forbs, had been completely excavated by feral pigs; the furrows of overturned soil were up to a foot and a half (45 cm) deep. The researchers described it as "entirely devoid of standing plants."

There is now evidence that an overabundance of pigs can change the makeup of a forest by determining which plants are able to grow and which ones get trampled, eaten or dug up. In a 10-year study done in Holland, researchers found that forests protected by fences from incursions of wild boars and deer took on a dramatically different appearance. While unfenced plots were dominated by Scotch pine and beech, the fenced plots sprouted predominantly oaks and birch trees. Recent studies in the Pasoh

Reserve suggest that pigs may be having a similar effect on the vegetation in this important remnant of Malaysian rain forest.

But that's just one way in which exploding pig populations may be altering the environment. Ecologists are worried that too much rooting can affect the health of a forest by increasing soil erosion, disturbing the roots or damaging the bark of mature trees, increasing the rate at which nutrients are leached from the soil, fouling supplies of fresh water and destroying habitat used by ground-nesting birds, shrews, voles and salamanders. Invasive pigs have also caused problems as predators, particularly on islands where native species lack instincts for defending themselves. On Santiago Island in the Galapagos archipelago, egg predation by feral pigs is believed to have reduced populations of beach-nesting giant tortoises and sea turtles. In Tennessee's Great Smoky Mountains, feral pigs are threatening rare native salamanders and snails.

The environmental impacts of feral pigs, like those of humans, are believed to have larger ecological repercussions. By disturbing native vegetation, for example, pigs open the door to invasive plant species. And they can alter food webs. On California's Santa Cruz Island, for example, the spread of feral pigs in the 1990s is believed to have restructured the top half of the island's food web. Traditionally the diminutive island fox, a rare species that feeds on spotted skunks and other prey, had occupied the role of top predator. Once pigs became common, however, golden eagles began venturing over from the mainland and eventually set up permanent residence. The trouble was that, even though they had more than enough piglets to eat, the growing eagle population also preyed on foxes. By the turn of the millennium there were signs that the foxes were in serious trouble, and the island's nocturnal skunk population, freed from the threat of its main predator, exploded.

And ecological damage is just the half of it. Feral pigs are now barging into the human world in the same reckless way they tear up a rain forest. And humans are increasingly not amused. Farmers suffer the most. In Europe wild boars eat maize, grains, potatoes and sugar beets, to name a few. In Indonesia they raid coconut plantations. In Pakistan the targets are sugarcane and peanuts. In northern Australia they rampage through

In Germany, both the population and territory of the feral pig have increased dramatically and the pigs are now a serious traffic hazard. Automobile collisions with pigs have resulted in numerous human fatalities around the world.

sugarcane fields and banana plantations. In the United States pigs devour grain crops and corn, and in the wine-producing states they're particularly fond of grapes. As if all this weren't enough, feral pigs will also attack sheep during lambing season. A recent study from Luxembourg quantified the resulting costs from 1997 to 2006, finding that wild boars had caused over US$7 million (5.2 million euros) in damage in that one country alone. Often the problem isn't so much what the pigs eat as the careless way they do their eating. According to another study, no more than 5 to 10 percent of the damage caused by pigs represents food that was eaten; everything else is costs associated with smashed fences and trampled plants.

While crop loss is nothing that an electric fence can't minimize (albeit at considerable expense), other problems are harder to deal with. Hog farmers, for example, must make efforts to keep their animals free from certain infectious microbes that can flare up unexpectedly and decimate a herd. Feral pigs act as carriers for many of these agents, and a single midnight raid is all it takes to contaminate the entire farm. Even more worrisome, feral pigs may have a capacity to short-circuit the safety lines established to protect the human food supply. When spinach contaminated with a dangerous strain of *E. coli* killed three people and sickened more than 200 others across North America in the fall of 2006, investigating officials eventually singled out pigs as the likely cause of the problem. Feral pigs were known to have been present on the farms in California where the tainted spinach originated. It's thought they may have contaminated the crop after picking up the *E. coli* while wandering through feces on nearby cattle ranches.

Off the farm, out-of-control feral pig populations cause even more problems. In recent years pigs have begun raiding suburban residential gardens, yards and garbage in a growing number of places, sometimes with unnerving aggressiveness. In April 2009, a 26-year-old woman in St. Petersburg, Florida, was reportedly attacked by a 200-pound (91 kg) wild pig that had wandered into her backyard. In some areas pigs have joined deer as dangerous traffic hazards; according to one recent estimate, road-crossing feral pigs are now responsible for 27,000 auto accidents in the U.S. each year.

Gradually humans are beginning to fight back. The ecological damage associated with these invaders has driven conservationists to organize several large-scale pig slaughters. On Santiago Island in the Galapagos, nearly 30 years' worth of hunting and poisoning — an effort that resulted in the death of more than 18,000 pigs — finally came to an end in 2000, when the invaders were declared officially extirpated. Similar campaigns have been waged successfully on the islands of Santa Cruz and Catalina, off the coast of California. In the latter case, just under US$3.2 million and more than 96,000 hours' worth of effort were needed to rid the island of 12,000 pigs. The conservationists killed pigs in most every way one could imagine: with dogs, from the sky, at night by spotlight, on the ground by hand gun, and with traps.

Success in these cases was of course aided by the fact that the introduced populations lived on islands. On continents it's a different story. Indeed, efforts to date indicate that even massive ongoing culls aren't enough to keep pig populations under control. In recent years the number of pigs culled under direction of the U.S. Department of Agriculture's Wildlife Services has climbed from 5,351 across 10 states in 2001 to 13,423 across 19 states in 2005. In 2006 federal authorities killed more than 14,500 feral pigs in Texas alone. Despite this, the state's estimated population of roughly two million pigs continues to thrive. (While this same government agency is exterminating feral pigs, it is also very busy killing predators; in 2007 alone it oversaw culls of nearly 90,000 coyotes, roughly 4,700 foxes, 338 mountain lions and 340 gray wolves.)

One problem facing feral pig eradication efforts is that not everyone wants them eradicated. In places like Hawaii, where the animals have been present for centuries, feral pigs and feral pig hunting are now well established as an important part of local culture. Many people across the southern United States also value wild pigs for sport hunting and as a source of game meat; for landowners they can be a valuable resource for these same reasons. In such cases conservationists and wildlife biologists are now searching for ways to manage the situation so that feral pig numbers are low enough not to cause any major damage but still sufficient to provide value as a resource. Of course, attempting to exercise such massive levels of control over nature invariably requires time and effort, not to mention the risks of unforeseen consequences. In a world increasingly void of major predators, however, one has to wonder: do we have any other choice?

GIANT AFRICAN LAND SNAIL

Achatina fulica

WHEN IT COMES TO the giant African land snail, it's hard to know where to begin. Certainly there's the species' unnervingly large size and propensity for leaving behind a disgusting mess. Then there's its slow, seemingly inexorable march around the globe. But we must also take into account its recent links with the spread of human-disease-causing agents and its tendency to drive us to what can only be described as incredibly dumb actions — including launching biocontrol programs that stand among the biggest blunders in the history of humankind's efforts to control invasive species.

In the end it's maybe precisely that: the degree to which these big, slimy invertebrates can leave us not knowing what to think. Giant African land snails have been tremendously successful at living in a changing world, but is that really such a bad thing? They're big and ugly, but should we let that get to us? They eat more than 500 different crop species, but does that justify expensive, ecologically damaging eradication campaigns? Just about the only certainty is that in today's modern world, *Achatina fulica* is here to stay.

As their name suggests, giant African land snails are difficult to miss. They're often referred to as the world's largest land snails; the biggest specimen on record had a shell length of 8 inches (20 cm) and a total body

◀ *Achatina fulica* is rarely found in natural, undisturbed forests; the species thrives in human-modified environments and can even be found in big cities.

length of close to a foot (30.5 cm). Size aside, they are not unlike many other herbivorous land snails. They eat dead and decaying plant matter. They keep vampire hours, staying out of sight when the sun is high and emerging at dusk to forage throughout the night. They move slowly and not very far, laying down a trail of lubricating mucus as they go.

A certain amount of irony is associated with *A. fulica*'s rise to prominence. Snails and slugs are one of the most diverse yet one of the most threatened groups of animals on the planet. Scientists estimate that, besides 24,000 known species of land mollusk, possibly as many as 40,000 others have not yet been described. According to a 2009 study, 422 species are now extinct; this figure represents roughly 40 percent of all animal extinctions listed on the International Union for Conservation of Nature's 2009 Red List of Threatened Species, the *Who's Who* of species on the path toward extinction. And since so little is known about most of the remaining members of this group, the true figure is no doubt much higher.

By contrast, *A. fulica* is in the midst of an epic march that's among the most spectacular of all species invasions. Scientists aren't sure exactly when it began. There is evidence that the species, which is native to the coastal regions of East Africa between Kenya and Mozambique, may have been introduced onto the island of Madagascar in the 1700s. What they do know for sure is that over the past two centuries these snails have been moving forward — eastward, for the most part — in one long, remarkably consistent expansion. It's a diaspora that, if you were to depict it chronologically on a map of the world, would resemble fireworks — wave after wave of starbursts giving birth to more starbursts. It really needs to be laid out in detail to be appreciated. It began in the Indian Ocean, where giant African land snails reached the island of Mauritius, to the east of Madagascar, around 1800, then the Comoros Islands (precise date unknown), the Seychelles (around 1840) and the Indian subcontinent (1847), and into what is now Sri Lanka by the end of the century.

At this point the invasions moved forward on several fronts. In Southeast Asia the giant snails turned up in Malaysia in 1911, followed by Singapore (1917), Indonesia (sometime in the 1920s) and Borneo (1930), Thailand and Vietnam (around 1937) and Burma (the early 1940s). To the north they were first seen in the Philippines and on the Chinese mainland in 1931, in what is now Taiwan around 1932 and in Hong Kong in 1937. The South Pacific saw the next wave, from the Mariana Islands, the Bonin Islands of Japan, Hawaii and Palau during the 1930s to New Guinea in 1945, the Society Islands and Vanuatu in 1967, New Caledonia in 1972, the Marquesas Islands in 1979 and the Samoan Islands around 1978. During this time the snails were also proliferating through most of India, reaching into Bangladesh to the east and Pakistan to the west by the 1960s. In the Indian Ocean they continued to spread even further, invading the Andaman Islands (after 1939), Christmas Island (the 1940s) and the Maldive Islands (around 1957). They also began moving west on their native continent, making it to Morocco by 1983 and to the Ivory Coast (1988) and Ghana (1994) in West Africa.

At some point in this long march it became evident that it was only a matter of time before the snails would arrive in the New World. And sure enough, in 1984 they turned up in Guadeloupe, in the eastern Caribbean. Soon after that, giant snails were found on Martinique (in 1988), St. Lucia (2000), Barbados (2000) and Antigua (2008). In October 2008 the government of Trinidad and Tobago issued a press release stating that *A. fulica* had been spotted in the far northwestern corner of the island of Trinidad. South America hasn't been spared either. In 1988 the snails were being offered for sale at an agricultural fair in southern Brazil. By 2006 there had been recorded sightings in 23 of the country's 26 states.

Giant African land snails, despite what you may be thinking at this point, cannot fly. Nor can they swim. The question then is how an animal that moves literally as slowly as molasses can circumnavigate a mostly watery planet with a circumference of nearly 25,000 miles, even if it did take more than 200 years. Like most things *A. fulica*-related, it's a tough one. The short answer is — humans. The long answer, though, seems to have more to do with what one might be tempted to call biological destiny. One way or another, they seem capable of getting from one place to another, so a better question might be, is there anything that *hasn't* contributed to their spread?

A partial list begins with Mauritius, where giant African land snails are believed to have been imported in 1800 by a governor's wife suffering from chest pains. Historically, snails have been used as medicine for a number of ailments — sometimes just the mucus, other times the whole snail, crushed, boiled or puréed — and it was hoped that the mollusks would provide her with a cure. Unfortunately the poor woman died, and the unneeded snails were let loose. In some places such as India, the exotic appeal of giant snails led to their being imported for use as living decorations in botanical gardens. During the 20th century several introductions in Southeast Asia were the work of duck farmers who were interested in a self-renewing source of food for their birds. Other introductions were made by people who were interested in raising giant snails for either food or medicine. Many populations in the Pacific islands were established during the Second World War by Japanese troops, as an emergency food supply in case of invasion.

In the 1980s they were brought to Brazil as part of a trend that was taking the country by storm — snail farming. Media reporters and silver-tongued snail importers convinced hundreds of people to set up snail ranches to cash in on what was promised to be a lucrative business. But the whole thing was a complete flop for a number of reasons, including the fact that Brazilians, unlike the French, don't like eating snails. As one farm after another quickly went out of business, thousands of snails were left to fend for themselves.

Not all of *A. fulica*'s introductions have been intentional. Tourists have returned home with what they thought were empty shells when in fact they were carrying temporarily dormant but still perfectly healthy snails. On many other occasions snails or snail eggs have been unknowingly moved from one part of the world to another as part of shipments of plants, produce, nursery stock or soil.

A. fulica is not invincible. These animals are highly intolerant of arid conditions, and the door slams shut on them as soon as they venture into a climate zone with anything resembling a winter. Humans too can be a real pain. In places where money is not in short supply, the expansion of giant African land snails has been held at bay by a combination of vigilant customs inspections and timely implementation of control programs. This happened in Australia after a colony of giant snails was discovered in Queensland in 1977. And it's happened several times in the United States, including in California in the mid-1940s, in Arizona in 1959 and in Florida in 1966.

The Florida incident was a close call. The outbreak began when a Miami boy returned from Hawaii with some interesting shells, which his grandmother eventually put in her garden. Unknown to anyone, they contained live snails. The snails multiplied, and by the time they were discovered in 1969 they had infested most of the city, prompting authorities to launch an eradication campaign involving 128 tons (116 tonnes) of poison bait, pesticide spraying, weed and garbage removal, manual collection of snails, and production and distribution of more than 325,000 brochures aimed at generating help from local residents. In 1975, after almost five years of hard work and a cost of $700,000, the program was declared a success. In all, some 18,000 giant snails were found and destroyed.

But success stories have been few and far between. Usually giant snails have invaded without hindrance, and when this happens they've been known to rapidly build up large populations that often seem overwhelming. In one study from the Maldives, researchers estimated an average of seven snails per square foot (73 per sq m), and other studies found densities of between four and five per square foot (46–56 per sq m). While such estimates can be misleading because of the species' tendency to congregate in certain areas, some sense of the magnitude of the situation can be gleaned from the results of control programs. In 1931, three years after giant snails were first spotted in Sarawak in northwest Borneo, a bounty resulted in collection of some 500,000 snails and 20 million eggs over the course of just two weeks. On the island of Bugsuk in the Philippines, another eradication campaign reportedly collected 45 million snails over a 4,000-acre (1,600 ha) area during a seven-month period in the 1980s.

A. fulica has several interesting traits that help make it such a successful invader. Like other snails it's hermaphroditic, meaning that each one possesses both male and female reproductive organs and has the potential to lay eggs or impregnate. What's more, once a snail has mated — a slow-motion love scene that can last for 24 hours and result in both partners becoming pregnant — for a year and a half and perhaps even longer, it can carry on laying fertilized eggs without needing to mate again. Thus it is possible for a new population to spring from the introduction of a single snail.

Like many invasive species, giant land snails are also capable of rapid reproduction. They don't live very long — probably five or six years — and they don't lay more than a few hundred eggs at a time. They can, however, reach sexual maturity in less than half a year. And while many animals are restricted to annual or seasonal birthing schedules, when conditions are right, giant snails are capable of churning out batch after batch of eggs as frequently as every few weeks. The rest is simply mathematics. Albert Mead, the world's leading expert on *A. fulica*, studied the species for 60 years before his death in 2009, and he once put pen to paper to get a handle on the reproductive capacity of giant snails. He started with very conservative figures — an age of maturity of nine months and four batches of 150 eggs per year over a five-year lifespan — and figured out what sort of empire a single snail would be lording over at the end of a three-year period. His answer: nearly eight billion snails. Tack on another two years and the total number of progeny adds up to over 16 quadrillion. (Mead was evidently fascinated by the exercise, because he went on to calculate what that enormous figure would be in terms of distance: enough snail footage, he discovered, to get you to the sun and back 5,477 times.) By comparison, endangered species like the tree snails of Hawaii don't reach sexual maturity until around six years of age, and when they do they often produce only one offspring at a time.

Another feature — again, one that's common to many different types of land snails — is the ability to enter into a dormant state known as estivation. During periods when it is hot or dry, snails will seek some form of shelter by digging into the ground, crawling under a rock or even climbing up a tree, and then pull their bodies deep inside their shells. To prevent water loss, they secrete a special membrane, called an epiphragm, which temporarily seals off the shell's opening. Often giant snails will emerge from this state at the first sign of rain. They're capable of going for long periods without food, however — up to 10 months, according to one study — so theoretically estivation could easily outlast even a drought. This trait has made it easier for humans to unwittingly transfer giant snails around the world. When they're in hiding, they're easily overlooked. And when they're uninterested in moving about, they're in a good state to survive the hardships of long journeys. The snails' genetic programming to seek out every nook and cranny also makes it challenging to mount eradication campaigns whose success depends on finding every last individual.

Last and likely not least, *A. fulica* possesses that other suite of traits that almost invariably characterizes today's modern Super Species: hardiness, adaptability and the ability to thrive in habitats that have been modified in various ways by humans. Many of today's imperiled land snails are species whose lifestyles are specialized for certain types of forest conditions, and their fate is due largely to destruction of their habitat. *A. fulica* (and most of the 200 species belonging to its immediate family) is just the opposite. Time and again researchers have commented on how giant African land snails are almost completely absent from unmodified forests wherever they've invaded. Where they've been found in the greatest numbers, on the other hand, are those places that one might categorize as "unnatural" — farms, plantations, vegetable gardens, flower gardens, roadside clearings and garbage dumps. This includes not just agricultural or other rural settings but also backyard gardens in villages and even big cities.

Part of this adaptability has to do with a varied diet. Giant snails in their

Among the largest species of land snail, giant African land snails can grow to a length of close to 1 foot (30 cm) and achieve a weight of over 1 pound (450 g).

native environment are restricted to eating dead and decaying plant matter at a pace that makes them one of nature's fastest composting units. They must have iron guts, however, because when they enter the world of humans they'll eat almost anything — fruits, leaves, soil, dead animals and feces, as well as cardboard and other forms of garbage. The 500 plants that scientists have identified as being palatable to giant snails are testimony to this dietary range. Giant snails are also adept at blending into different habitats. They vary greatly in size, shape (from slender to slightly obese) and color (from a rich dark brown to almost white), and this likely helps them adapt to conditions found in different environments. Likewise, researchers have found that their daily habits and life cycles can also be adjusted to suit local conditions.

Not surprisingly, giant snails have long had a bad reputation. Between their voracious appetites and their tendency to undergo population explosions upon arrival in a new territory, these invaders long ago established themselves as serious agricultural pests. "Giant Crop-Devouring Snails Threaten U.S.," declared one headline in the *New York Sunday Times* in 1949. In addition to the economic damage associated with eating crops, there's also the nuisance factor. Consider, for example, the slime. Like all snails, *A. fulica* has a remarkable ability to generate large amounts of thick mucus. It's a wonder fluid whose secrets (the way it expands explosively upon first contact with air, the way it becomes slipperier the harder you rub it) researchers have been trying to unlock for decades.

At the peak of an invasion this slime is everywhere, along with snail bodies and snail shells. Mead described the situation this way: "The snails multiply in such unbelievable numbers that they crawl all over and into everything, they crush or slip out from under foot almost wherever one steps, they cover things with their excreta and sticky slime trails, and they die in great quantity for various reasons and create rank odors." A few years after the invasion in Florida, a state agriculture representative used almost identical language to describe the situation there. American soldiers landing on the South Pacific island of Saipan after the Second World War found themselves facing a most unusual hazard: Jeeps sliding out of control on roads whitewashed with snail mucus.

More recently, the name *A. fulica* has been popping up in relation to the spread of emerging human diseases, particularly a condition known as angiostrongyliasis. This disease is caused by a parasitic worm, *Angiostrongylus cantonensis,* that lives in the lungs of rats but can get into humans who eat certain intermediate hosts — snails, for instance — or vegetables that carry the worms in their larval stage. The disease has been found most frequently in Asia and the islands of the South Pacific, with a worldwide total of fewer than 3,000, mostly mild cases since it was discovered in Taiwan in 1945. In a human the larvae fail to develop into adult worms, but they do infiltrate human intestines and sometimes the central nervous system. The result can be severe headaches, neck stiffness, tingling sensations, vomiting, nausea and sometimes, if treatment isn't administered, death.

Attempts to exterminate giant snail populations have failed more often than they've succeeded. In Malaysia people tried collecting them and throwing them in the ocean, but many of the snails were able to make it back to shore unharmed, even after long periods of submersion (studies have shown it takes two full days of total submersion to drown giant snails). A better approach proved to be putting the snails in large sacks and then burying them deep underground. Early attempts at using chemical weapons were hampered by the fact that most first-generation agricultural pesticide systems were designed for killing insects. A mollusk, being of different biological makeup, requires its own form of poison. Eventually formulas containing copper sulfate or a combination of calcium arsenate and metaldehyde proved more effective, but only when regularly reapplied. In its war on giant snails, India has reportedly tried out 50 different molluscicides.

Such approaches are expensive and ultimately ineffective unless carried out on a massive scale and before populations have gotten out of control. Agencies in charge of pest control favor a much cheaper and far easier option known as biological control. With this approach the aim is to identify and release some form of organism — it could be a predator, a parasite or a disease-causing agent — that attacks and eventually controls or even eliminates a given pest. One of the most ambitious programs targeting *A. fulica* was launched in Hawaii beginning in the 1950s. Backed by powerful agricultural groups representing, among others, pineapple growers and sugar producers, scientists tested a list of organisms that included ants, beetles, flies, crustaceans, birds, various types of worms, mites and even single-celled protozoans. In the end the authorities reached a decision that seems mind-bogglingly reckless in retrospect: they would fight *A. fulica* by unleashing droves of nonnative carnivorous land snails. Yes, they would wage war on introduced snails by introducing more snails — 15 different species of them, to be precise.

The scheme, which took place between 1955 and 1959, now stands as one of the biggest blunders the war on invasive species has ever seen. Instead of munching away on giant snails, at least three of the introduced nonnative predators — particularly *Euglandina rosea* from Florida, commonly known as the rosy wolfsnail — instead acquired a taste for Hawaii's native snails. This might not have been too bad except for the fact that the native land snails of Hawaii make up one of the most interesting and threatened assemblages of land snails anywhere. The enormous number of forms (one estimate recently described it as 931 species, 332 subspecies and 198 varieties, most found nowhere else) has made the islands of Hawaii to mollusks what the Galapagos Islands are to finches — a model for studying how animals can evolve to inhabit different ecological niches. Unfortunately, many Hawaiian land snails are susceptible to habitat loss; an estimated 600 species have been lost during the past two centuries. Many scientists now believe that rosy wolfsnails played a major role in pushing many vulnerable species over the edge. Even more tragic, scientists have seen no evidence that the introduced snails had any impact on populations of *A. fulica*. As prominent invasive species biologist Daniel Simberloff once concluded, "This entire operation of biological control against *Achatina fulica* has been disastrous."

Unfortunately, the story doesn't end there. In the decades after predatory snails were used in Hawaii for biocontrol, island nations all over the Pacific followed suit. Rosy wolfsnails have now been introduced into places that include New Caledonia, the Society Islands (including Tahiti and Moorea), American Samoa and Vanuatu. Eight different predatory snail species have also been introduced into India. No one knows exactly how much of a toll these schemes have taken, largely because most areas haven't been closely studied. However, there are fears that what's happening on Hawaii isn't unique. On Moorea the introduction of *E. rosea* came while an international team of researchers was studying evolution among snails belonging to the genus *Partula*. At the time of the introduction, in 1984, the island was home to seven species belonging to this group. By 1988 all seven had vanished. Across the entire Society Islands archipelago, 56 of 61 *Partula* snail species are now thought to be extinct. As if this wasn't enough, there's also evidence that another species introduced for biocontrol against giant snails — the predatory flatworm *Platydemus manokwari* — has played a role in the decline of native snails on the island of Guam, and possibly elsewhere.

It would be one thing if *A. fulica* were the extreme threat it's commonly regarded to be, but this assumption may not be completely accurate. For one thing, the full extent of the damage caused by these snails has never been carefully assessed. Some early studies concluded that these snails could be present in large numbers in certain places without causing much damage at all. In 1961 Albert Mead considered that much of the giant snail's reputation was due to media hype. "It is only in the most localized areas, under unusually favorable conditions," he wrote, "that damage by these snails will approach anything near the absolute." Eighteen years of investigation later, he hadn't changed his mind: "Even with the great numbers characteristic of young populations the damage is fairly localized, and not catastrophic or devastating on a broad scale."

Other scientists have argued that Mead's assessments failed to consider the fact that *A. fulica* invasions in some places have made it impossible to grow certain crops. Even so, there exists no tally of the actual cost of the problem. Instead there is well-documented proof that the massive population explosions of giant snails are usually only temporary. With time, *A. fulica* populations level off after their initial bursts, and they usually decline eventually to more tolerable levels. (Scientists aren't exactly sure why, but one theory is that as snail populations reach a certain density, they become increasingly susceptible to the spread of deadly snail diseases. They're limited, in other words, by their own success.)

At the same time, there is little evidence that *A. fulica* is a serious threat when it comes to the spread of human disease. Giant snails do carry rat lungworm, but so do many other snail species, as well as shrimp and other crustaceans. What's more, if *A. fulica* were a major source of the problem one would expect to see an increase in disease rates corresponding to a rise in the spread of the snails. No such evidence exists. In places like New Caledonia, giant snail explosions have even coincided with a fall in the number of meningitis cases.

In the end perhaps the key to better understanding *A. fulica*'s place in the modern world lies in a better quantitative assessment of its true cost. If the snails really are the serious pests that some people assume they are, then perhaps there's some way to justify the high cost that's been paid to keep them under control. But let's see some honest numbers. If, on the other hand, it turns out that *A. fulica*'s reputation is largely a myth perpetuated by media hype and self-interested agribusiness, then it is time to acknowledge that what's happened is a true tragedy.

C. DIFFICILE

Clostridium difficile

T HE TENDENCY FOR SOME PLANTS and animals to get out of control in response to environmental disturbance can be seen as a universal principal of ecology. It can happen in lakes where algae blooms knock back resident species, on grasslands where species such as cheatgrass proliferate in response to overgrazing, and in vacant urban lots or roadway clearings where kudzu and other super-weeds run wild over land cleared of its native vegetation. And it can happen in places not so readily associated with the onslaught of invasive species — like your own bowels.

Clostridium difficile is its name. It may be that you've never heard of it. It's a small, rod-shaped bacterium, one of the 15,000 to 36,000 species of bacteria that can take up permanent residence in the human gastrointestinal tract. Under normal conditions this diverse community of microbes, which also includes fungi, yeast, protozoa and as many as 1,200 different types of viruses, doesn't cause any problems. Indeed, in recent years they've come to be seen as an integral part of the biology of all multicelled life forms, helping us digest nutrients, maintain the right pH levels, fend off harmful outside germs, stimulate our immune systems and ensure proper development and maintenance of certain tissues and organs. Without them, we probably wouldn't be able to survive.

Although most details about your gut bugs remain largely unknown — making this netherworld one of the last frontiers waiting to be explored — studies done on the microbial jungles of the mouth reveal that the residents of the body, like species in a rain forest or alpine lake, live together

A rod-shaped microbe, *C. difficile* may be found in the ecosystem of the human gastrointestinal tract. If the human habitat is healthy, *C. difficile* does not cause problems, but if the human habitat has been altered by antibiotic use, an infection of *C. difficile* can be fatal (colored transmission electron micrograph [TEM], magnification X 15,500).

in a complex web of relationships that bears all the characteristics of an ecosystem. When we're born, we start out as newly created islands that are colonized within minutes by pioneering microbes. During our first year their complexity and diversity increase and pass through various stages of succession, in which one level of equilibrium gives way to another. Eventually we're left with intestinal old growth — a teeming pit of microbial diversity that's been estimated to include anywhere from 100 billion to 100 trillion individual organisms.

As with any other habitat, the environments of the body are thought to harbor all the typical ecological players. There are predators and prey.

The diarrhea that's associated with this disease, coupled with the fact *C. difficile* is a spore-producing bacteria, means the microbe is easily and readily transmitted throughout the environment. What makes the situation even more potentially explosive is the fact that the microbe is also known to persist for long periods of time and to resist being killed by common disinfectants. When researchers at the Cleveland Veterans Affairs Medical Center recently conducted a study to see where *C. difficile* might be lurking, they discovered that it was widespread: in both nursing stations and physicians' work stations, including on telephone keypads, desktop computers, tabletops, doorknobs, sinks and various pieces of portable medical equipment. In all, 16 percent of patient rooms and 23 percent of surfaces were found to harbor *C. difficile*. In a home setting, where patients are likely to encounter only healthy people, this normally wouldn't cause any problems. In hospitals, particularly intensive care units, where there are likely to be many people on antibiotics, it can be a major problem. In some outbreaks, up to 40 percent of patients have become infected, with about one-third developing colitis.

An even bigger problem, meanwhile, is that by creating new breeding grounds for *C. difficile*, we also create opportunities for evolution of new and potentially more dangerous strains. In recent years an increasing number of people are getting sick from *C. difficile*, and the percentage of patients that end up dying is also growing. Both trends have been blamed on the emergence of a new super-strain.

Researchers are only now trying to understand how different species in the invisible world of the intestines work together to form a stable, well-balanced ecosystem. One goal is to identify specific species that are particularly important in keeping *C. difficile* in check under normal circumstances. Such species might then be used in drugs that would be taken along with antibiotics, the aim being to help restore the gut's rich ecosystem as quickly as possible and in the process keep the invasive *C. difficile* at bay.

Part Three

ARRIVAL OF
THE FITTEST

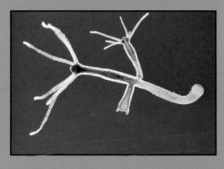

INVASIVE SPECIES HAVE SPREAD so frequently and in so many places that it's easy to underestimate the magnitude of their achievements. By one commonly used rule of thumb, only 10 percent of the species that find their way into foreign habitats are able to establish self-sustaining populations. Of these, only one in ten are able to spread widely enough to earn recognition as a true invasive species. Thus, out of every hundred species that are given a chance to invade new territory, only one is able to take advantage of the opportunity.

What is it about these species that makes them super-invaders? One prominent idea, known as the enemy-release hypothesis, is that invasive species undergo rapid population explosions in new environments because they've left behind their natural predators, parasites and disease. Plunked down in a new environment, they have an unfair advantage because native species are still battling their long-standing enemies. This argument is one reason why many people think of invasive species as something like an attack by space monsters: strange organisms arrive from out of the blue, kill everything in their path and then take over the world. It's also the logic underpinning the belief that invasive species are a direct threat to native biodiversity.

But evidence for the enemy-release hypothesis remains open to debate. For one thing, studies have shown that invaders are frequently subject to predation and disease in their new environments. Sometimes even more diseases plague the alien species than the natives, as was found in studies done on invasive grasses. Second, there's no good explanation for how life-forms could so easily leave their enemies behind, particularly disease-causing microbes. Finally, the fact that invasion occurs with only a small percentage of introduced species suggests that the secret to success involves far more than just moving from one place to another. Clearly there must be something about the species themselves, some degree of what invasion biologists now refer to as "invasiveness."

According to evolutionary theory the answer boils down to a single word: fitness. Possibly one of the most misunderstood ideas associated with evolution, fitness can be viewed as the ability of an organism to successfully reproduce in the face of both competition from other organisms and the challenges imposed by its physical environment. Fitness measures the combined effects of the countless biological characteristics that make up any life-form — traits associated with acquiring and utilizing resources, defending against predators and microbes, surviving the stresses caused by cold, heat and dryness, and so on. Fitness can also be viewed as a by-product of evolution. If a random genetic mutation gives an organism advantages that lead to enhanced output and survival of offspring, then that variation can be said to have improved the organism's fitness. Thus, over time, organisms become adapted to the relatively more stable aspects of their surroundings.

It's not possible to determine fitness simply by analyzing an organism's biological traits. This is because fitness stems not from the traits themselves but from how useful they are under a given set of environmental circumstances. It's nature's nasty curveball: a trait that may contribute to the success of a species in one environment may be detrimental to the same species in a different setting. Likewise, an organism may be the dominant species in its environment one minute and then, if conditions within that environment shift — as they often do — that same species may find itself on the path toward extinction.

These ideas suggest that two separate forces lie behind the emergence of today's Super Species. On the one hand, by subjecting habitats to an unprecedented level of upheaval, humans have altered the conditions that helped shaped the evolution of many native species. These formerly successful species now find themselves struggling to survive under conditions to which they are not well adapted. At the same time, these altered environments set up competitions among species to see which are most suited to life under the new order. The winners are those species now taking over the globe — Super Species that have what it takes to thrive in landscapes created by humans.

Not surprisingly, many of the world's super-invaders possess adaptations that give them a competitive edge in newly emergent environments such as cities or in heavily disturbed habitats such as polluted lakes. Among plants, for instance, the perfect "weed" is a species that can reproduce rapidly (usually both sexually and asexually), grow quickly from seedling to sexual maturity and be able to survive under different conditions, through what biologists call phenotypic plasticity — the ability to call on backup traits that enhance survival under different conditions. Cheatgrass, an Old World plant that has taken over much of the North American plains, is a prime example. In addition to reproducing via both cross-pollination and self-pollination (which frees it from dependency on any particular insect while still allowing for the advantages of genetic mixing that come with cross-breeding), it can flower at different times and thrive in areas with widely varying soil depth, pH levels, precipitation and nutrient availability.

Invasive insects, meanwhile, tend to be species whose life cycles move at hyper-speed. They're small, even for insects. They grow quickly, reaching sexual maturity at an early age, and multiple generations can come and go in a single season. Invasive birds have also been found to share certain traits. They're often aggressive in pursuit of both food and nest sites. They reproduce rapidly, laying relatively large numbers of eggs, often multiple times a year. And they are frequently bold, which enables them to tolerate the presence of humans and to venture into unfamiliar environments. Many highly invasive fish and mammals also have rapid reproduction rates relative to other species from their groups, as well as the ability to acquire, eat and digest a range of different foods. They also commonly possess a tendency for some individuals to migrate away from established populations. And, no less important, they frequently have characteristics that appeal in some way to humans. Trout, for example, are for the most part lousy invaders, yet they're present in lakes worldwide, mainly because sport anglers put them there year after year.

The world is also being taken over by generalists. These are species that can survive in a wide variety of environmental conditions, get nutrition from lots of different sources and adapt well to novel situations. In short, their survival does not depend on the health of a single host species, one type of food or a particular habitat feature. This distinction — which involves a graduated scale of variation rather than one clear dividing line — may reflect how different habitats have helped drive evolution in certain directions. Many generalists trace their origins to regions that have undergone repeated periods of environmental flux caused by natural forces, such as flooding or changes in salinity.

Species that carve out highly specialized ecological niches tend to do so in highly competitive ecosystems with a long history of relative stability. Among ants, for example, researchers have found that very sophisticated colonies — those with multiple queens and numerous specialized forms of workers — are extremely competitive in predictable environments. They are, however, sensitive to disturbance. Not surprisingly, these highly evolved species are not among the several ant species that have emerged as Super Species. Those that are spreading widely lack many of the behavioral bells and whistles that normally make ants such fascinating subjects of study. They aren't adapted for survival under desert-like conditions or in places with harsh winters. They're mostly just bold, aggressive, small and fecund — in other words, fearless and capable pioneers.

Such observations help provide a more accurate explanation for the current threats facing biodiversity. As habitats undergo destruction or modification, native species that have acquired habitat-specific adaptations find themselves either unable to survive or, at the very least, unable to compete with newcomers. This is particularly true if the newcomer has broader tolerances or adaptations that turn out to be useful in human-modified environments. Geckos are a good example. As visitors to tropical Pacific islands have discovered, these nocturnal insect-eating lizards are a common sight around human habitations. Before the Second World War one of the most common species was the native mourning gecko (*Lepidodactylus lugubris*). During recent decades, however, in many areas mourning geckos have been gradually replaced by the common house gecko (*Hemidactylus frenatus*), a species whose skill at wandering aboard sea vessels in port has helped it establish populations in Florida, Central America, Venezuela and Australia and throughout the warmer regions of the Pacific, including Hawaii. What was at first surprising about this displacement was the fact that female mourning geckos can produce offspring from unfertilized eggs, while house geckos must rely exclusively on sexual reproduction. This means that, theoretically at least, populations of the former should be able to expand twice as fast — a huge advantage, one would think, in the competition for survival.

During the 1990s researchers from the University of California at San

Diego conducted studies inside aircraft hangars at a military base on the Hawaiian island of Oahu. By adjusting conditions inside the otherwise identical hangars, the scientists investigated why house geckos thrive so consistently around buildings and why native gecko populations fail when the alien species is present. Since house geckos are larger and faster, the initial suspicion was that it was because of bullying. But even though the smaller lizards seemed to shy away from the invaders, there were no indications of any form of aggression. Further study suggested that it all boils down to how house geckos use their speed and size to advantage while hunting insects. For example, house geckos are often the first to attack whenever prey is present. They also attack from farther away, and when they do, they usually scare off other insects nearby. Mourning geckos are often left on the sidelines like the runts of a litter, waiting for hunting opportunities that rarely come.

The most important discovery of this research was the fact that the outcome may depend entirely on the nature of the habitat. The wall of a building is a particular habitat with a distinctive structure. Most obviously it's flat and featureless, two characteristics that may seem unimportant but actually mean a great deal in the battle between predator and prey. For one thing, there's nowhere for a hunter to hide. In addition, walls are frequently lighted, which acts as a lure to insects, concentrating them in a relatively small area. The scientists discovered that the house gecko simply has better physical tools for exploiting these circumstances. Its size and speed combine to help it overcome the featureless terrain. Meanwhile, the drawback associated with being big — the need for lots of food — is negated by the abundance of insects attracted to the light. One would expect, then, that the smaller native geckos would have the advantage amid the leaves and branches of a forest, a habitat where speed isn't so important and where insects are spread out and harder to find. And this may indeed be the case; native geckos continue to thrive in native forests and other habitats outside urban and suburban environments.

Of course, before an invasive species can even begin to test itself in a foreign environment, it has to get there. This means that in addition to traits useful for invasion, a successful Super Species must have traits that enable it to use humans for dispersal. Some characteristics are useful in both regards. Species that aren't afraid of humans, for example, tend to venture into packing crates and other places where they're likely to get shipped overseas. Species hardy enough to handle varying environmental conditions are also more likely to survive weeks or months inside the ballast water of a ship, attached to a boat being towed cross-country, or stuffed inside a tourist's suitcase. Other traits may be useful only during this initial stage of colonization. For example, self-fertilization in plants and the ability of inseminated female animals to store sperm for extended periods (common among insects, reptiles, birds and bats) make it theoretically possible for a single individual to establish a new colony of the species.

Successful dispersal is also heavily dependent on the prevailing winds of human behavior. Larger species are generally more useful to humans — usually as food sources, but also to control pests or, in the case of trees, for erosion control and landscaping — and until recently such species were frequently spread around the world intentionally. But now that rising invasion-related problems have curtailed deliberate introductions, size has become an impediment. If you're a goat or a mongoose — two early examples of intentional introductions gone awry — the chances of your being accidentally introduced anywhere are now pretty much zero.

While these general rules provide a solid theoretical framework for understanding invasiveness, conservationists are frustrated that it's still not possible to identify which wandering species might emerge as the next super-invader. Some species seem to have all the attributes but have so far shown only limited invasive potential. At the same time, not all successful invaders possess all of the key traits. This may be due in part to the fact that every invaded habitat represents a unique situation: each is a unique ecosystem shaped by its own unique set of conditions. With so many variables to consider, it could be that biological mathematicians simply haven't yet developed computer modeling tools sophisticated enough to accurately predict the outcomes of species invasions in specific environments.

What complicates things even more is that invasion success may be due not only to traits possessed by a species before invasion but also to those that may be acquired after it has become established in a new territory. In an intriguing new insight into invasion biology, researchers are finding evidence that introduced species may be capable of rapidly acquiring useful adaptations while the invasion is in progress, either as a consequence of the invasion process in general or possibly because invaders have an inherent capacity for hyper-evolution. The acquisition of resistance to pesticides and other control measures is the best-known example. But now there's evidence that invasive species are capable of undergoing more general transformations that presumably boost fitness under their new circumstances. The generalists, in other words, are finding their niches in the world of humans.

One good example involves St. John's wort, a medicinal herb that, since its introduction during the mid-1800s, has become widespread as an invasive species in a range of habitats throughout North America. Recent experiments have shown that specimens from different parts of the plant's invasive range grow better under their respective conditions: northern specimens outgrow southern ones in northern test plots, and vice versa. This could be because the invaders originated from different habitats, but genetic analyses suggest that the differentiation has taken place since the start of the invasion.

A similar conclusion emerged from studies done on reed canarygrass, a Eurasian wetland resident that spread throughout eastern North America after being repeatedly introduced for forage production, sewage treatment and other reasons during the mid-19th century. In a series of experiments reported in 2007, University of Vermont botanists Sébastien Lavergne and Jane Molofsky compared the DNA and biology of reed canarygrass taken from different parts of its native and nonnative ranges. The results show that North American strains are like Frankenstein versions of their native ancestors, with high degrees of genetic diversity and mixing of DNA — likely the result of sexual reproduction that occurred in North America between various strains of widely different native origins. Not only that, greenhouse experiments done by the same researchers show that North American versions outperform their European cousins in several areas related to invasion success. The former emerged from seeds faster with a higher rate of successful emergence, produced new shoots faster, had higher rates of leaf production and produced more overall plant matter.

Meanwhile, a similar phenomenon of change may be at work in animals. At least this is the conclusion one might draw from the observation that cane toads appear to have evolved longer legs and greater endurance since their introduction into Australia — a possible response to the advantages of being able to spread into new territory at a faster rate.

ARGENTINE ANT

Linepithema humile

FOR FOUR WEEKS during the spring of 2000 a team of European entomologists embarked on a collecting mission that took them along much of the Atlantic and Mediterranean coasts of Europe. The purpose was to identify the colony boundaries of the Argentine ant, *Linepithema humile*, a highly invasive species that is gradually taking over a good portion of Earth. Roughly 165,000 worker ants from 33 populations were tested to determine their degree of relatedness. In one such test, workers from different populations were placed together in a vial. If they fought — as competing ants almost always do — it was a sign they belonged to unrelated colonies. If they got along, it was an indication of close kinship.

The results were, to say the least, surprising. Three of the populations appeared to form one large supercolony extending several hundred miles along the Mediterranean coast of eastern Spain. Even more astonishingly, all of the remaining ants belonged to a second supercolony that was even bigger, a massive constituency of ants that blankets territory stretching more than 3,700 miles (6,000 km), from the Atlantic coast of northwestern Spain to northern Italy. The results, supported by genetic analysis and other experiments, suggest that southern Europe has been overrun by an enormous mob of nearly identical, like-minded insects; it consists of millions of nests and billions of individuals. If this is what Argentine ants

◀ Argentine ants are quick to initiate fights with other insects and will battle to the death. Although they are not especially successful fighting one-on-one, they excel at team combat and will gang up to overpower competitors.

141

are capable of, the researchers concluded, it's no wonder they're taking over the world.

Ironically, *L. humile* looks anything but intimidating. These ants, which can be light to dark brown in color, are quite tiny as far as ants go, with full-grown workers reaching barely more than 0.1 inch (3 mm) in length — less than the thickness of a shirt button. The ant world is full of physical traits that evolved in the name of biological struggle — poisonous stingers, enlarged mandibles, terrifying castes of enlarged soldiers to brutally intimidate rivals — but Argentine ants seem positively bland by comparison. In their native habitat, the forest floors around big rivers in northern Argentina, Paraguay, Uruguay and southern Brazil, they are just another species struggling to survive in a highly diverse and competitive ecosystem. Indeed, when researchers visit these areas, Argentine ants are sometimes difficult to find. Yet despite all this, *L. humile* has become possibly the most successful ant species on the planet.

The first invasion reports date back to the late 1800s, initially around the Portuguese wine region of Madeira and shortly thereafter in the southeastern United States, near New Orleans. Since then they've become established throughout the southern U.S. from the Carolinas to Texas, as well as in much of southern California and scattered sections of Oregon and Washington. In Europe the species has taken over most of the Mediterranean regions of Spain, Portugal, France and Italy; additional invasions have been reported in Great Britain and Switzerland. In South America they have expanded their range into Chile, Ecuador, Peru and previously uninhabited portions of Argentina and Brazil, and they've also moved into parts of Central America and Mexico. Elsewhere, Argentine ants have been seen in Australia since the 1930s and in New Zealand since 1990. They've also become established in Japan, the United Arab Emirates and South Africa. Some 35 countries have been invaded so far, a number that is likely still growing.

On one level this spectacular advance can be viewed as a human-driven phenomenon. Under their own steam, Argentine ants are ill-equipped for long-distance dispersal. For one thing, they can't cover a lot a territory very quickly. Nor do they survive very well outside their preferred habitat

— areas that are neither too hot nor too cold and that have plenty of year-round moisture. This means that the species would have been hard-pressed to move much beyond its native range in South America, let alone skip between continents, without some form of help.

Like many other invasive species, Argentine ants have gotten a boost from human trade. In this case it's believed they've been transported around the world as stowaways, either along with plants destined for nurseries or in shipments containing soil or other agricultural products. Once the escapees arrive on foreign soil, a single queen and as few as 10 workers may be all that's required to found a new colony. At any rate, such introductions appear to be far from rare, judging from the presence of Argentine ants not only on six different continents but also in faraway places such as Easter Island and Hawaii.

Human-driven alteration of the landscape also seems to be fueling the spread. In their South American river habitat, Argentine ants evolved a semi-nomadic, highly adaptable lifestyle to survive an environment frequently turned upside down by flooding. While some ant species tend to stay put, tending the same nest for long periods, Argentine ants readily establish new nests in response to discovery of new food or other changes in their environment. They're also adaptable when it comes to where they nest. As a result, they are at ease in a range of human landscapes and human-modified habitats, from farms and suburbs to inner-city parking lots and inside homes and other buildings.

Perhaps even more important, Argentine ants are most at home under conditions that humans consider ideal. They bask in sunny Mediterranean climates with enough moisture to sustain sap-sucking insects that are attracted to succulent plants. Normally places like southern California and the hills of Tuscany would be too dry, but watering our suburban gardens and irrigating acre after acre of farmland have created an Argentine ant paradise. And it's no coincidence that the vast majority of nonnative areas where Argentine ants are now found are habitats that have been heavily modified by humans. Indeed, some researchers think that human modification of the land may be a prerequisite for the Argentine ant's ability to outcompete native species. As researchers for a study in southern

California recently concluded, "Our results suggest that anthropogenic [human] modifications to the physical environment are preeminent in determining the extent to which mediterranean scrub communities in southern California are susceptible to invasion by Argentine ants."

There is, however, one problem with viewing this invasion as entirely a human-made phenomenon. With the intensification of global trade, many of the world's roughly 12,000 known ant species are being transported around the planet. Although the number of invasion opportunities is great — in the U.S. alone border officials detected 232 different foreign ant species between 1927 and 1985 — only a handful of species come close to matching the current success of the Argentine ant. Clearly these ants are outcompeting potential rivals in the scramble for ecological footholds in a changing world. And researchers are slowly getting to the bottom of what makes this remarkable species so special.

One factor may be the Argentine ant's diet. In 2007 a group of researchers led by Andrew Suarez at the University of Illinois published the results of a study in which they tracked the progressive westward expansion of Argentine ants in Rice Canyon, a narrow treed corridor running through the heart of densely populated southern California. One of the surprising results of the eight-year study was how the ants appeared to thrive on two completely different diets. The researchers discovered that during the initial stages of the invasion, Argentine ants were voracious carnivores, feasting almost entirely on the protein-rich blood of native ants and perhaps other insects. Over time — likely because they had literally eaten their way through the native populations — this gave way to a diet mainly of carbohydrates, in the form of honeydew stolen from plant-eating insects such as aphids. Thus, even though Argentine ants are known to be carnivorous in their home environment, they're equally adept at thriving on sugar if needed. Such dietary flexibility, the researchers conclude, helps explain why Argentine ants can thrive in a variety of habitats, particularly disturbed habitats with reduced ecological diversity — two likely keys to their success as invaders.

But there's more to it than that. Argentine ants are also extremely aggressive. At Purdue University in Indiana, entomologists Grzegorz

Buczkowski and Gary Bennett recently designed a series of experiments to test the competitive abilities of Argentine ants against one of North America's peskiest ant species, the odorous house ant (*Tapinoma sessile*). The researchers found that when single workers from each species were put together, the result was almost always a confrontation marked by brutal violence. The two ants would engage in physical aggression, unleash potent chemical toxins or sometimes rely on both tactics to win ugly battles that often did not end until one of the combatants was dead. Although Argentine ants were not incredibly successful warriors (they won only 45 percent of the confrontations), they were more than twice as likely to be the ant that started the fight in the first place. They also worked more successfully as members of a team. When the researchers pitted groups of 20 workers against each other, the Argentine ants' aggression turned the fight into a rout. They quickly ganged up on individual house ants in a series of battles that almost always led to total domination. The Argentine

Instead of competing against one another, Argentine ant colonies tend to work together, forming massive "supercolonies." The ants above cooperate to move a larva.

ant forces suffered huge losses, but the house ants fared even worse: they had lost an average of 95 percent of their workers by the end of the eight-hour experiment.

Other tests by the same researchers further demonstrated the ferocity of Argentine ants. When lone workers were plunked down in the vicinity of a house-ant nest, the Argentine ant still went on the attack, even though the result was always a foregone conclusion. But this willingness to go boldly forth paid off in group situations. For example, when two separated colonies were given simultaneous access to a new food source, it was hardly even a contest. The Argentine ants found the food first, set up new nests near the food and even in enemy territory, monopolized the food and brutally attacked and slaughtered their opponents. In six of eight such experiments, most of the similarly sized house ants were killed. In the other two, they were wiped out completely. In all these group experiments the researchers noticed a striking propensity among the Argentine ants to work together. While house ants almost always engaged in one-on-one attacks, the Argentine ants invariably ganged up on individual house-ant workers, sometimes outnumbering them six to one. While one or two immobilized the hapless victim, additional workers got busy methodically snipping it into pieces. As Australian researchers recently concluded following a similar study, Argentine ants "express pronounced physical aggression above levels normally occurring between competing ants." The more Argentine ants there are, in other words, the more successful they'll be.

Enter the supercolony phenomenon. In their native range Argentine ants are like most other species of ants: tight-knit families of perhaps several thousand individuals engaged in ongoing battles with neighboring nests. These competing nests include different ant species as well as unrelated colonies of Argentine ants. In many places where they've invaded, however, Argentine ants appear to develop into massive groups of genetically similar individuals spread out over large areas. The supercolonies consist of multiple nests, and the main difference is compatibility — instead of fighting one another, neighboring colonies cooperate.

Supercolonies have now been identified in Australia, New Zealand, Japan, Bermuda, Chile and North America. In southern California, researchers identified a supercolony that stretches from Ukiah, north of San Francisco, to San Diego — a distance of more than 560 miles (900 km). The largest colony recorded to date, however, is the one in southern Europe. Covering most of the coastal regions of Spain, Portugal, France and northern Italy, it consists of billions of closely related ants spread over a mostly narrow strip running parallel to the Atlantic and Mediterranean coasts. And the supercolony situation may be even more unsettling than first thought. In the summer of 2009 Japanese researchers discovered that Argentine ants from California, southern Europe and Japan refuse to fight one another, a finding that suggests they're part of one enormous megacolony now overrunning the world. In addition to providing the species with the ability to directly crush any competitors, large populations give Argentine ants other advantages in the struggle for ecological dominance. Several studies have shown that the ants are not only quicker to find new food resources, they're also able to recruit additional foragers sooner and in greater numbers. As a result they completely dominate exploitation of existing sources of nutrition.

Not surprisingly, success has made Argentine ants a bit of a nuisance. Economically they've been blamed for damaging crops such as oranges, figs and avocados by feeding on blossoms and fruit and by spreading diseases between plants. By protecting insects that produce honeydew, they're also effectively bolstering the populations of pests that feed on and destroy a wide range of other crops. Homeowners are not particularly fond of these intruders either. Argentine ants can usually find some way of getting into houses and other structures, particularly in warm climates, where careful sealing of windows is not a common practice. This, coupled with the species' adept foraging skills, means that anything sweet left exposed overnight will invariably be swarming with workers by morning. Besides the annoyance caused by long trails of intruders and spoiled food, Argentine ants that nest inside buildings have been known to damage walls, insulation and electrical wiring.

Ecologists are also concerned. Studies have found that Argentine ants have displaced native species in California, Hawaii, Japan and South Africa by up to 90 percent, and there is evidence that local ecosystems are being altered as a result. In many areas of southern California, for example, Argentine ants have displaced harvester ants, which are much larger insects that survive by transporting seeds and other organic matter into permanent nests dug deep in the ground. The loss of harvester ants is thought to be altering characteristics of the soil in these areas by removing this important source of aeration and enrichment. Their decline might also be affecting insectivores such as the coastal horned lizard (*Phrynosoma coronatum*). Research indicates that these reptiles, already imperiled by loss of habitat, simply can't get enough nutrition from eating the much smaller Argentine ants.

On the other hand, the fact that Argentine ants are moving into disturbed habitats suggests they may be taking advantage of ecosystems that were already in peril, in which case they are not the cause of the problem but a symptom. There's even evidence suggesting that Argentine ants may be playing a positive role. Studies have shown that some areas now have a higher total ant biomass thanks to the arrival of *L. humile*. In Portugal, Argentine ants have been shown to limit the spread of the pine processionary moth (*Thaumetopoea pityocampa*), a pest whose larvae can destroy trees if left unchecked.

Either way there seems to be little that can be done to halt the invasion. Once present, Argentine ants are difficult to control and almost impossible to eradicate, and so far researchers have had little success devising a strategy for bringing them under control. The only hope seems to be nature's seemingly inevitable tendency toward self-correction. Indeed, another invasive species, the fire ant, is now challenging Argentine ants in some areas. Also, scientists versed in the laws of genetics and evolution predict that the supercolony phenomenon will eventually undermine itself because of the rise of cheaters — workers that prosper by mooching off their nest-mates' efforts. But that's conjecture. To date the advance of the Argentine ant shows no sign of abating, and projections of global warming have led some to suggest that these tiny terrors will soon have plenty of new territory to conquer, including large areas in Africa and Southeast Asia. The invasion, in other words, may have only just begun.

CROWN-OF-THORNS STARFISH

Acanthaster planci

CORAL REEFS WERE ONCE viewed as the ocean's version of equatorial rain forests — biologically rich habitats that epitomize ecological stability. Then along came *Acanthaster planci*, the crown-of-thorns starfish. A brutish creature, as starfish go, *A. planci* was considered rare until 1962, when large numbers were first seen off the coast of northeastern Australia. Then almost overnight the species began multiplying so successfully that some sections of reef were carpeted with crown-of-thorns starfish — piled two deep in places.

While the mob gradually expanded southward along much of the Great Barrier Reef, similar eruptions were observed on reefs throughout the Pacific and Indian oceans. And everywhere these outbreaks occurred, the devastation was clear: expanses of colorful coral transformed into bone white deserts that soon turned gray under a film of algae and other encrusting organisms. The carnage was such that some marine biologists wondered aloud whether reef systems could survive such an onslaught. Journalists, ever eager to embrace a possible apocalypse, proclaimed that the entire Great Barrier Reef was in danger of being destroyed.

In the mid-1970s concern melted away — the starfish vanished almost as quickly as they had appeared. But the celebration didn't last for long. A second, even worse eruption occurred between 1979 and 1991, followed by a third that lasted through the 1990s until the first signs of waning appeared in 2008. Today scientists remain dumbfounded. How is one species able to inject so much chaos into an ecosystem that's supposedly dedicated to

order? More important, are the outbreaks normal, or are they a sign that the wheels of the reef ecosystem are about to come flying off?

Certainly *A. planci* possesses many qualities that would seem to qualify it for Super Species status. As one of the largest members of the starfish family, it usually reaches the diameter of a dinner plate, but divers have encountered some specimens that seem to be from another planet — great, ungainly beasts close to 3 feet (1 m) in diameter that lumber around with the help of between 8 and 21 arms. From what scientists can tell, few predators find adult starfish attractive. They come equipped with formidable body armor that consists of a dense mat of spikes as long as matchsticks. These spines are sharp, and they're also highly venomous. Beachcombers unlucky enough to step on one or incautious scientists who prick their fingers may find themselves suffering from severe pain and swelling and a need to vomit that can last for days. When threatened, a crown-of-thorns starfish will curl itself into an ominous spine-studded ball.

Such traits are thought to greatly aid *A. planci*'s ability to withstand potential predators. Indeed, while juveniles are known to avoid predation by feeding at night and hiding among the nooks and crannies of the reef, adults take advantage of their formidable armor to feed all day long out in the open. Even if a fish or some other carnivore takes a bite, these starfish have another factor to contribute to their confidence: like other members of their family they're capable of regenerating lost limbs.

Another advantage *A. planci* has is its ability to eat coral polyps. Polyps are the animals responsible for building a reef. They secrete calcium carbonate to form a hard skeleton; when the polyp dies, this skeleton, along with those of countless neighbors, becomes part of the hard coral that eventually creates a reef. Corals have been able to cope with their predators partly because they store large amounts of a waxy substance, cetyl palmitate, that makes them extremely difficult to digest. For the crown-of-thorns starfish, however, this isn't a problem. Their stomachs, unlike those of most other starfish, release large quantities of special enzymes that are highly efficient at digesting wax. Thus, while other starfish scavenge the reef for algae, bits

◀ A diver spears *A. Planci* in attempts to remove the species from a coral reef.

of organic debris and various critters such as sponges and shellfish that can't flee, crown-of-thorns starfish can exploit the virtually limitless bounty of living coral with hardly any competition.

They do this with considerable dexterity. Again unlike most of its kin, *A. planci* has a soft, flexible underside thought to have evolved so that young starfish can wrap themselves around the delicate tips of coral branches, where most of the living polyps are found. A crown-of-thorns starfish also has a specialized feeding mode that some may find slightly creepy: it can get a jump on the digestion process by unfurling a portion of its stomach through its mouth. Once this gut tissue is draped over a polyp patch, the starfish secretes potent digestive juices that dissolve its meal into a liquid that can more easily be slurped up and absorbed. When this is accomplished, it's time to haul in the stomach and head for the next patch of coral. Each year a single starfish can graze its way through as much as 65 square feet (6 sq m) of coral — not bad for a creature whose top speed is 0.01 miles an hour.

Having had nearly five decades to study the biology of the species, scientists are convinced that it can't do much harm when its numbers are low. For one thing, when there are few coral-eating competitors about, it seems to focus on certain types of coral, many of which are fast-growing and thus capable of rapid regeneration. In such cases the coral's growth rate is able to outpace the grazing effects of the starfish, and reef building is unaffected. When outbreaks occur, however, the choosiness disappears. Crown-of-thorns starfish will attack and destroy any coral they can get at, including slow-growing species whose individual structures may have taken centuries to form. Eventually the point is reached where the coral is being eaten faster than it can replace itself.

Reef surveys indicate that *A. planci* attacks tend to be patchy, an indication that the species may rely on prevailing currents to get from one section of coral to another, as well as from one reef to another. Areas that do get hit during an outbreak, however, get hit hard. In the worst areas, over 90 percent of the living coral has been destroyed, with the numbers of starfish on individual reefs often reaching the hundreds of thousands, and sometimes even millions. No other single threat — not cyclones, extensive

The starfish's trademark "thorns" cover its body, and while they are only about as long as matchsticks in most specimens, they are extremely sharp and highly poisonous. One prick can render a human victim ill for several days.

attention from humans, pollution or damaging warm spells — can generate an impact so severe. At this point there is nothing left for the starfish to eat and they are thought to undergo rapid population loss, likely due to disease outbreaks triggered by overpopulation.

Scientists remain baffled by why these outbreaks are occurring. On the one hand there's a possibility that this may be a natural part of reef ecology. Humans, after all, have been exploring marine environments only since the invention of diving gear during the Second World War. For all today's marine biologists know, crown-of-thorns outbreaks may have occurred on reefs ever since reefs first appeared. Indeed, an analysis of bits of fossilized spine buried in the ocean floor indicated that the species has been chowing down on Great Barrier Reef coral for at least 8,000 years. More controversially, some scientists think this same study provides evidence for ancient outbreaks. This interpretation, if correct, would support the idea that periodic outbreaks, perhaps driven by the El Niño — southern oscillation ocean/atmosphere phenomenon, or fluctuations in temperature, salinity or plankton levels — are normal. Furthermore, ecologists generally think that populations of species able to produce large numbers of offspring in a short time are inherently unstable. On this count, the crown-of-thorns starfish certainly fits the bill. When conditions are right, a single female can produce up to 60 million eggs each year.

But another study, in which researchers examined feeding scars on long-lived corals, provides evidence that outbreaks have become increasingly frequent in recent decades. This supports the view that such outbreaks are not natural, but rather a side effect of the environmental changes induced during the past century by humans. It has been proposed, for example, that overfishing and overharvesting of reef resources such as the giant triton snail and its 20-inch (50 cm) shell — the traditional trumpet of choice among South Sea islanders — has resulted in reduction of predators that would formerly have kept the crown-of-thorns under check by preying on juveniles. This theory lacks direct evidence, and there is little proof that triton snails or any other predator eats huge numbers of starfish. But recent surveys have revealed that among the Australian reef areas most favored by *A. planci*, those protected by anti-harvesting laws are nearly four times less likely to be hit during an outbreak. Some researchers now believe that overfishing of top predators can have a dramatic cascading effect on the food chain. In this case, commercially harvested fish such as coral trout normally eat the smaller fish that in turn feed on various invertebrates whose diet would include juvenile starfish. Fewer trout would mean an increase in smaller fish, which would result in fewer starfish predators and thus more starfish.

Finally there's pollution. It has been observed that many outbreaks around the Pacific have begun two or three years — the time it takes for starfish to develop to a point where they can be easily detected — after heavy rains, particularly heavy rains that followed a prolonged dry spell. They also occur frequently near human settlements. This has led some researchers to propose that agricultural and urban runoff temporarily alters the near-shore ocean environment in a way that boosts the survival rates of young starfish. Exactly how isn't clear. It's been suggested that pesticides flushed into the ocean by rain may have less impact on *A. planci* than on its

Another reason for their success is the fact that, despite their relatively large size, nutrias breed like house mice. Blame this on the species' impressively efficient reproductive biology: Female nutrias can breed continually and can fit almost three pregnancy cycles into a single year, sometimes producing up to a dozen young in a single litter. The offspring are born fully furred, with their eyes open. They're able to feed on their own within hours of being born and continue to grow at amazing speed. And in barely no time at all — six months in the case of females — the young are ready to start churning out babies of their own. Under good conditions, mother nutrias are capable of producing an average of 15 offspring a year. With an average lifespan estimated to be around six years, it's not unlikely that a long-lived mother could give birth to a hundred offspring over the course of a lifetime. That, as they might say down in the bayous, is a lot of swamp rats.

But there are likely additional explanations for the nutria's success, including the state of ecological chaos in a large portion of the world's wetland habitats. The expansion of human populations has had a major impact on these environments. River valleys have been dammed and converted into farmland. Saltwater marshes and inland swamps have been drained and sold as prime real estate. Nonnative marsh plants have overrun native species. At the same time, pollutants flushed from waterside cities and towns; enormous quantities of fertilizer, pesticides and other chemicals seeping off farms; and livestock overgrazing are all having a major impact on wetland ecology throughout the world.

Some species such as the muskrat have not responded well to these changes, but nutrias seem unfazed by the new face of Earth. They thrive in fields of sugarcane and rice. They readily raid backyard gardens. They show surprising tolerance for a degree of muck and filth that most species would find repulsive. Perhaps the best example comes from the cattle lagoons of Florida. During the early 1970s, researchers investigating the arrival of nutrias in the Sunshine State quickly discovered that the animals were drawn to areas not normally listed in the tourist brochures. As the authors of one scientific paper noted, "It became evident that nutria were exceedingly abundant in barnyard runoff canals and polluted holding ponds maintained by several large dairies in the Tampa Bay area." These sites included what were essentially sewage treatment ponds — pools full of bacteria where all the manure generated each day by a farm is dumped. The nitrogen-rich dung meant that the ponds — as well other waterways close to farms — supported thriving masses of shoreline vegetation and water plants, including highly invasive nonnatives such as water hyacinth, which some farmers have planted to help speed the water purification process.

To nutrias, such conditions are evidently quite tolerable. One manure pit near Tampa was found to have a nutria density of 25 animals per hectare (about 10 per acre), four times more than an otherwise regular (i.e., non-poop-filled) pond on a nearby farm — among the highest densities of nutria ever recorded anywhere. Such hardiness under extreme environmental conditions led the study's chief scientist, University of South Florida biologist Larry Brown, to correctly predict a future world filled with nutria. Wrote Brown in 1975, "This species thus exhibits the potential of being as ubiquitous in certain disrupted aquatic systems as is the black rat (*Rattus rattus*) under more terrestrial situations in many portions of the world."

In addition to adaptability, one additional factor — one that can also be blamed on humans — appears to be helping nutrias succeed: an absence of predators. It takes more than a small snake or an owl to bring down a giant rodent, and in modern times large predators are in increasingly short supply. In many wetland areas of the southern United States, for example, extensive alligator hunting has left nutria with virtually no predators except humans. Populations of many nutria predators elsewhere in the world — jungle cats, jaguars, cougars, ocelots and golden jackals, to name a few — have also been greatly reduced by habitat destruction and overhunting, especially in and around developed or otherwise disturbed areas.

The one nutria predator that does exist in great numbers — humans — also happens to be a fickle one. During must of the second half of the 20th century, high fur prices encouraged many trappers to pursue nutria on a large scale. By the 1980s the number trapped each year in Louisiana alone had reached one million. The species was once seen as a valuable resource, but when wearing dead animal pelts went out of fashion in the late 1980s, prices began to fall, trapping declined and nutria populations exploded.

The nutria's strange physiology makes it well adapted to life in the wetlands; females have nipples located high on their sides to allow their young to nurse on land as well as in the water.

The environmental impact of so many large plant-eaters can be enormous. By digging into the sides of embankments, nutrias can disrupt the integrity of levees and other water-containment systems. As the hosts of parasites and microbes, they can spread disease. And by clear-cutting cordgrass and other water plants down to their roots, they can trigger tidal expulsion of nutrient-rich sediments, which are needed to sustain not just plant life but also mussels and the many other organisms drawn to thick marshland vegetation that make such habitats so biologically rich. As nutria numbers have grown, such impacts have shifted from being minor nuisances to major concerns; the valuable resource has become an obnoxious pest.

Efforts to control nutria invasions by poisoning, bounty hunting and other methods have met with limited success. While the species has been successfully eradicated by extensive campaigns in Great Britain and California, it remains solidly entrenched and seemingly impervious to removal efforts in places like the southern United States — a new wetland resident that appears here to stay. Today there's a whiff of irony rising from the nutria-filled swamps. On the one hand, expensive campaigns are being waged to control the invasion of nonnative plants such as water hyacinth, which crowd out native species and make waterways difficult for boats. At the same time a concerted effort is being aimed at eradicating nutria because they forage extensively on water plants — including water hyacinth.

State authorities in Louisiana believe they have a solution that may help restore some ecological balance to their nutria-ravaged wetlands: they are encouraging us to eat our way out of the problem. To that end the Department of Wildlife and Fisheries recently challenged chef Philippe Parola to come up with interesting and tasty new recipes featuring nutria. The resulting dishes include nutria chili, nutria hind saddle with mustard sauce, nutria sausage jambalaya and nutria fettuccini. If that's not exotic enough for you, how about *ragondin à l'orange*?

Whether nutria becomes a human staple or not, there's at least one major lesson to be learned from this invasion, and it's that large-scale ecosystem disruptions have major consequences. Once you do something as drastic as reducing the top predators, there's nothing to keep voracious herbivores from eating all the plants. In an ideal world humans would have figured out how to live side by side with predators large enough to bring down large, agile rodents like the nutria. But this would require preservation of large areas of habitat suitable to the needs of such predators — a major change in our stewardship of the planet, and a hurdle we've yet to figure out how to jump.

HYDROZOAN

Turritopsis dohrnii

WE HAVE BEFORE US two rather disturbing developments. The first stems from a paper published in the journal *Experimental Gerontology* in 1998, "Mortality Patterns Suggest Lack of Senescence in Hydra." In it, Daniel E. Martinez of Pomona College in California describes his efforts to observe whether a small group of primitive predatory animals known as hydras experienced senescence, a fancy name for what the rest of us call aging. His conclusion? "Hydra may have indeed escaped senescence and may be potentially immortal."

The second turn of events is the recent concern among some scientists that ocean ecosystems may be on the verge of major changes because of rising sea temperatures, pollution and the accelerating rate at which humans have been harvesting marine resources. In a nutshell, their hypothesis states that by drastically changing the nature of the oceans, we're conducting a massive ecological experiment. The outcome may be marine environments dominated by the ocean's more primitive predators, namely jellyfish and their close cousins, which include the hydrozoans, the animal class that includes hydra.

The convergence of these ideas forces one to ponder a couple of equally unsettling questions. How long will it be before we see an invasive species emerge that combines a capacity for capitalizing on marine ecosystems that are off-balance with the ability to live forever? More to the point, is it possible that we're already witnessing such an invasion, in the form of one *Turritopsis dohrnii*?

It's okay if you haven't heard of this little animalcule; even scientists are just beginning to come to grips with its unusual life cycle. It's hardly more than a speck, an insignificant little creature whose delicate beauty hides its true nature as a ruthless predator of plankton. It has recently made headlines after reports that, in addition to cheating death, it also has the ability to reverse its development, essentially turning itself back into a preadolescent after having reached adulthood. Oh, and there's also new evidence that suggests it may be carrying out what one researcher termed "a silent invasion."

To begin with, hydrozoans are about as weird as they come. Officially classified as animals, they eat single-celled protozoa, water fleas, tiny shrimp, insect larvae and other minute creatures. Yet many species form colonies that are attached to the ocean floor and look more like sea plants. In some cases they will release tiny buds that become free-floating individuals, known as medusae, that eat, swim and regenerate, through sexual reproduction, the next generation of colony-forming polyps.

Odder still are the mobile hydrozoans that resemble single entities but in fact are colonies of separate individuals. The Portuguese man-of-war is one example. These large creatures start out as a single plant-like polyp that, through a budding process driven by asexual reproduction, eventually develops into four separate polyps, each with its own shape and biological function. Even though the four are separate bodies, they can't survive on their own. Instead they remain tightly bound, functioning as a single unit that resembles a large jellyfish.

Whether or not some hydrozoans also possess immortality — at least in the sense of not being forced to make the death march known as aging — has been under debate for more than a century. The key is the observation that every cell in a hydra's primitive body undergoes continual renewal. Normally the various tissues of multicellular animals are made from cells that stop dividing once their initial development is complete. Over time such cells accumulate defects and toxins that inhibit function, and this is

Hydrozoans are able to reproduce asexually, through a process of budding. ▶
The strange life cycle of some species may also include the ability to
"cheat death," which the creature can do by reversing the aging process.

thought to be a contributing factor in the process of aging. In hydras, by contrast, stem cells are continually pumping out replacements for all the animal's specialized tissues — it's thought that no cell is ever more than 20 days old. From this, some researchers have hypothesized that under ideal conditions — no predators, no lethal environmental stressors — there's no reason why a hydra would ever stop living.

Of course forever, as they say, is a long time. Nonetheless, scientists attempting to identify signs of aging in hydras — checking to see, for example, if death is a greater probability among older animals — have been unsuccessful. Martinez's work in the early 1990s was more telling. In his lab he was able to maintain four separate hydra populations for four years, charting their progress along the way. A graph in his 1998 paper, comparing his results with known mortality curves for worms, fruit flies, guppies and voles, is revealing. While populations of the other species begin experiencing a sharp rise in deaths the older they got — culminating in 100 percent mortality long before four years has elapsed — the lines representing death rates among the hydra are dramatically flat. Almost none of them died.

The second shocker from the weird world of hydrozoans comes from earlier work done by a team of European researchers on *T. dohrnii* (mistakenly identified as *T. nutricula*). Previously the free-floating buds produced by hydrozoans had been seen as a one-way street: the medusae were freed, they lived, they produced and released either sperm or eggs, and then they disintegrated. But among the *Turritopsis*, researchers witnessed a surprising reversal. In a variation of the usual life cycle, medusae subjected to food shortages or changes in water temperature were seen to settle into the substrate and revert back to the plant-like colony state of their earlier developmental stage — a feat that at the cellular level would require complete retooling of most of the organism's body. This ability to reverse the direction of a developmental pathway suggests that these hydrozoans have evolved a potential way to escape death indefinitely.

Since those results appeared in 1996, researchers have found evidence of reverse development in other hydrozoans and their relatives. This has led to the theory that it is a widespread phenomenon among these primitive creatures, and has also rekindled the debate over whether it's possible for any single organism to live forever. On the one hand, of course, the question seems entirely academic. An organism may live forever theoretically, but in practice the presence of predators and other realities of the natural world make such a scenario impossible. Writing in the *Canadian Journal of Zoology* in 2004, a team of Italian researchers commented: "By implication, they could attain immortality. But if this potential for avoiding death were expressed in the field, these animals would saturate the world's oceans! Instead, we see that they suffer mortality, and population growth is kept under control."

Then again, maybe this isn't always the case. In a paper titled "A Silent Invasion," which appeared in the journal *Biological Invasions* in 2009, marine biologist Maria Pia Miglietta details her discovery that *T. dohrnii* appears to be undergoing a range expansion of global proportions. Miglietta analyzed DNA sequences from different *Turritopsis* species collected in various ocean locations around the world. While most species were found in only a single part of the world, sequences belonging to *T. dohrnii* turned up everywhere she looked: Japan, both the Atlantic and Pacific coasts of Panama, Florida, Spain, Italy. A lack of variation between the various *T. dohrnii* samples, meanwhile, suggests that the invasion was a recent event.

While it's still much too early to tell how common these tiny creatures really are, what factors might be driving their spread, or how far it's destined to go, the peculiar nature of the species' reversible life cycle and potential immortality makes this a compelling new chapter in the record of invasive species. Like many marine aliens, it seems likely that *T. dohrnii*'s spread may be aided by ballast water that cargo ships take on at one end of their transoceanic journey and then discharge at the other. Miglietta speculates that having the ability to turn itself young again in response to stress makes *T. dohrnii* ideally suited for this type of long-distance hitchhiking. However, she also notes that at least one of the sample sites is 185 miles (300 km) from the nearest harbor or shipping lane. Writes Miglietta, "The invading trajectory of *T. dohrnii* is thus expanding beyond the main ship traffic routes." A disturbing thought indeed.

KUDZU

Pueraria montana

THERE IS PERHAPS no single image of an invasive species that's more awe-inspiring, more evocative of a scene from a B-movie horror flick, than that of an abandoned lot overrun by kudzu vines. Picture an old, boarded-up house with a porch. Outside there's a rusty car, some shrubs and a tree or two lining the edge of a yard; on the outskirts, a telephone pole and maybe a mailbox. Now imagine the entire scene — every square inch of every surface — hidden beneath a thick, billowy blanket of leaves, obscured as completely as if the landscape had been hit by an all-night blizzard. Except for the contours, there is nothing to hint at what lies beneath — only a rolling sea of luxuriant green.

Scenes such as these, as unreal as they may sound to many, are actually commonplace in states such as Georgia, Alabama, Mississippi and South Carolina, where the invader's long list of colorful nicknames includes foot-a-night vine, the vine that ate the South, the green menace, the cuss-you vine and the mile-a-minute vine. Across this region, kudzu and humans are locked in an intense turf war that's been raging for decades. So far there's no end in sight. Nor is it clear which party will emerge victorious. As a result, a plant that under any other circumstances would be seen as a gardener's dream has now become a problem weed around the world, from Australia to Italy to South Africa.

◀ A parasitic weed, kudzu often uses trees to "climb" toward sunlight. Kudzu can similarly take advantage of human-made structures — like telephone poles.

169

Kudzu — known to scientists as *Pueraria montana* — actually has a long and illustrious history in human culture, particularly throughout its native range, which includes a wide swath of southern Asia. In China, for example, kudzu has been an important part of traditional medicine for more than two millennia. It has been used to treat dysentery, flu, fever and the effects of snake and insect bites; it's also been touted as a remedy for hangovers. Even today one can search the Internet and find kudzu root powder, kudzu root extract and other products offered as remedies for these and other ailments.

In addition to its medicinal uses, kudzu's sinewy stems have long been valued as a raw material for fashioning rope, baskets and other crafts. It's adorned gardens and homes as an ornamental plant. And its leaves and shoots have been a popular forage crop for cows, goats and sheep, providing a nutrient level on par with alfalfa and certain types of hay. Humans can eat it too. Indeed, historians have argued that this hardy vine, because it often grows when and where most other plants cannot, may have helped people survive droughts and famine.

One thing that made kudzu so popular in the past is the vine's hardiness. Although it thrives in direct sunlight, it can also survive in partial or full shade. It is viewed as drought-tolerant and it can persist in soils that are harsh, barren or otherwise inhospitable. In the mid-1950s this hardiness was put to the test on a patch of heavily polluted land in the southern Appalachian Mountains. Decades earlier sulfur dioxide emissions from copper smelters had killed off the native forest over about 10 square miles (28 sq km) of landscape. The topsoil had subsequently eroded away, leaving a lifeless wasteland. Several initial attempts to replant the area — with various grasses, legumes or young pine trees — all failed. Kudzu, on the other hand, took root with only modest amounts of fertilizer. When scientists visited the site a decade later, they noted that where the vine had been planted, the soils were teeming with insects, microbes and other soil-enriching organisms. By comparison, the neighboring non-kudzu areas were still lifeless. (This study of kudzu's ability to grow under such circumstances was partly due to the American government's interest in what organisms could be counted on to re-vegetate and restore landscapes in the aftermath of all-out nuclear war.)

Several factors contribute to kudzu's overall fitness. One is the fact that, as a vine, it enjoys a major advantage over other forms of vegetation in one of the plant world's most important competitive arenas: the battle over energy-giving sunlight. When plants first made their way onto land around 400 million years ago, they enjoyed unlimited access to the sun. But as primeval real estate became scarce it is thought that an evolutionary arms race began among species, which ultimately led to the evolution of trees. By being able to make woody trunks and sturdy branches, these organisms had the structural support needed to literally rise up above all the other plants. In doing so they were able to completely monopolize available solar energy. For the species below the canopy, that left three main choices: take to life in the shade, stick to places where trees couldn't grow, or find some way to use trees to their advantage. Twining vines such as kudzu have taken this last route. By relying on trees for structural support, they're able to gain access to the sun without having to invest all the energy that's needed to produce massive trunks and branches. They are, in essence, parasites.

There are important advantages to such a strategy. While trees are investing large amounts of nutrients and energy producing trunks and branches, vines can focus on making leaves and roots. This is particularly true for kudzu. Researchers have found that leaves make up no more than 2 percent of the dry mass of a mature deciduous tree, but they count for as much as 28 percent of a kudzu vine. Kudzu's starch-laden root systems, meanwhile, are equally impressive. Both primary and secondary roots are capable of fast growth; over time a single vine can develop a system that extends 12 feet (3.7 m) down into the soil and weighs up to 400 pounds (180 kg) — as much as 50 percent of the plant's total weight. Such formidable roots help kudzu find soil nutrients and store water, not to mention making it extremely difficult to kill.

Two other keys to kudzu's success are also worth mentioning. First, like other members of the pea and bean family, kudzu has evolved a symbiotic relationship with certain root bacteria that enables it to fix nitrogen in the soil. This means that rather than having to rely solely on the (rare) presence of pre-existing nitrogen compounds — compounds like the ammonia used in many fertilizers — kudzu can exploit the vast reserves of simple molecular nitrogen that make up close to 80 percent of Earth's atmosphere. The second trick is its ability to rapidly adjust the direction of its leaves in response to changes in the intensity and angle of sunlight. This is thought to help the plant minimize water loss from evaporation during periods of intense sunshine and to maximize photosynthesis when light levels are low. There is even evidence that kudzu angles its outer leaves so that light can filter down to its deeper layers.

Taken together, these and other factors help explain kudzu's phenomenal ability to engulf a landscape. Unlike weeds such as dandelions that proliferate mainly through dispersal of seeds, kudzu's main mode of growth is by creeping. Individual vines grow longer at the ends, and they also produce nodes along the length of each stem. These nodes are capable of not only sending out side shoots but also putting down roots, eventually splitting off to form independent plants. The stems, which can be close to 200 feet (60 m) long, form dense thickets. Kudzu's rate of growth is also astonishing. Each vine in a thicket is capable of adding a foot (30 cm) a

Kudzu is capable of growing up to a foot per day and can increase in length nearly 100 feet (30 m) in a single growing season. Creeping kudzu is not stopped by obstacles in its path and simply grows over anything it encounters.

day during the spring and summer growing seasons, for a seasonal total of close to 100 feet (30 m). As a result, the race for space with other plants isn't even close to being fair. Kudzu often overwhelms trees and shrubs with a layer of leaves and stems that can be more than 8 feet (2.4 m) deep. The underlying vegetation, starved of sunlight, quickly dies.

Despite all this, kudzu is not overly invasive in its native range, and it's thought to be more or less under control in most of the areas where it has been introduced. This includes Japan, where it was brought from China in the 18th century, as well as Italy, Switzerland and South Africa and parts of northeastern Australia, Puerto Rico and Oregon. Only in the American south has the species gone truly berserk. To understand why, we must turn the clock back to 1876. It was in that year that kudzu made its first North American appearance, displayed as an ornamental plant in the Japanese pavilion at the Philadelphia Centennial Exhibition. Not long after, mail-order catalogs began advertising kudzu as an ideal ornamental. Its rapid growth and hardiness made it an excellent choice for homeowners looking to provide shade for porches in sun-scorched areas like the South. As an added bonus, its large, luxuriant leaves were pleasant to look at and its magenta flowers were both attractive and fragrant.

Demand for the vine further increased in the early 1900s, when kudzu came into widespread use as a forage crop for livestock. But its real peak of popularity didn't come until the Great Depression. That's when the U.S. government addressed the problem of widespread soil erosion that was plaguing much of the farm belt after years of deforestation, intensive farming and drought. It saw the fast-growing kudzu and its extensive root system as the perfect way to keep soil in place. In addition to distributing 85 million kudzu seedlings, the government went so far as to offer payments to landowners willing to plant kudzu on their land. It's been estimated that by the mid-1940s, kudzu had been planted over nearly three million acres (1.2 million ha). This was supplemented by the efforts of the Civilian Conservation Corps, which directed widespread planting of kudzu on publicly owned land. On top of all this, kudzu was promoted both as a source of starch and as a raw material for making paper and even clothing.

But the vine proved too successful for its own good and attitudes quickly changed. In 1953 the U.S. Department of Agriculture removed kudzu from its list of plants suitable for ground cover. By 1970 the vine had achieved official status as a weed; in 1997 it made the Federal Noxious Weed List, a designation that among other things places restrictions on the sale and transfer of plants between states and requires establishment of programs to control its spread on public lands. Kudzu is now thought to occupy somewhere between two and seven million acres (800,000–2.8 million ha), and counting. Recently the vine has moved into New York, Pennsylvania and Ohio, and with the advent of global warming there are now concerns that it won't be long before it finds its way into Canada. The International Union for Conservation of Nature (IUCN) includes kudzu on its list of the 100 most destructive invasive species, and in countless news reports it has been a source of sensational headlines and unrestrained warnings. One reporter called kudzu "the worst weed in the world."

It is often suggested that kudzu's success as an invader resulted from the vine's leaving behind its natural predators and diseases. Scientists, however, have identified many insect and bacterial pests that eat away at its leaves, prey on its seeds and even kill adult plants. A more likely explanation is that kudzu's spread reflects the unique circumstances that surrounded its release — circumstances that relate not just to the vine's biology but also to human social and cultural trends of the past century. The vine's failure to run wild in other parts of the world where it has been introduced further suggests that factors in addition to rapid growth and superior competitiveness are involved.

To be sure, much of kudzu's astonishing success in the southern U.S. is due to the region's warm summer temperatures, abundant rainfall and mild winters. This may be one of the main reasons why the vine's northern spread has been limited in the past, and why some expect it to move further north in the event of global warming. As if on cue, Canada joined the ranks of kudzu-invaded nations for the first time in the fall of 2009, after the weed was found blanketing a 1.5-acre (0.6 ha) patch of land in southern Ontario. But a second and likely even more important factor is kudzu's ability to thrive in a variety of human-modified landscapes. Kudzu can grow in the shade, but its true ecological niche is out in the open beneath full sun, either in clearings or along the edges of forests. This plus its rapid growth rate makes kudzu well suited for dominating areas where pre-existing vegetation has been disturbed. Under natural circumstances this would include patches of forest where trees have been knocked over by storms. Far more common today are places where humans have modified previously forested habitats with chainsaws and bulldozers — land that's been cleared for logging, to create farms and cities, or to make corridors for roads and railway and power lines. Indeed, one might say that in the past 200 years or so most of the eastern United States has been converted into one giant kudzu farm.

Some historians argue that the proliferation of kudzu in the American south is also intricately linked to the social and cultural events that marked the first half of the 20th century. With the collapse of cotton farming and the movement of Americans away from farms and small towns and into bigger urban centers, there was a widespread shift in land-use practices throughout much of the South. As part of this trend, many areas where kudzu had been planted and maintained were abandoned or converted to woodlands. This allowed the vine to run wild, spread rapidly and take over open spaces wherever control measures were absent. And as Southerners are all too painfully aware, the longer a kudzu vine is allowed to proliferate untamed, the harder it is to eradicate.

Kudzu is often painted as a threat to forests, native species and the health of ecosystems, but such opinions are based largely on fear and misunderstanding of the species' ecological niche. Although it can survive in shade, there are no well-known cases of kudzu invading, let along destroying, large stands of mature forest. Its preference for disturbed areas, meanwhile, means that its impacts are most heavily felt where native ecosystems have already been severely altered by the presence of humans. Even the degree to which it's a pest depends largely on your perspective. The harshest criticisms come from those who see the vine as an impediment to their efforts to exploit the land. Loggers who clear-cut forests are issuing an open invitation to kudzu, and unless they invest heavily in costly control measures, a new crop of trees simply won't stand a chance. No farmer in his or her right mind would leave a kudzu infestation unattended.

But others see the kudzu invasion in a more positive light. Like many people in China, more and more Americans are beginning to explore the vine's potential as a resource. Kudzu is known to contain chemical compounds with medicinal properties, including daidzein, which is used to fight inflammation and infections, and daidzin, which is believed to possess anti-cancer properties. Various artisans have developed unique styles for making baskets and paper from kudzu. Others have been trying to revive the use of kudzu as forage for sheep and goats. There is even a strong push to position kudzu as a raw material for producing bio-fuel, with research suggesting that this weed could come close to rivaling the yields of current ethanol sources such as corn.

Last but not least, kudzu enthusiasts see the vine as an inexpensive and readily available source of nutrition. Entrepreneurs are now experimenting with kudzu-blossom jelly and kudzu-blossom syrup, kudzu hay, kudzu tea, kudzu salad mix and stewed kudzu root. The starch in the ground-up root makes it an excellent substitute for thickening agents such as cornstarch and flour. Others have come up with a wide array of recipes featuring kudzu leaves sautéed, steamed, stuffed, boiled and even deep-fried with a little garlic. Taken as a whole, one could say that the kudzu invasion has sparked a cultural awakening in the form of shifting resource use. Given the vine's regenerative potential and hardiness, perhaps the only valid question is why it's taken so long.

Part Four

THE WORLD
OF THE FUTURE

IN THE WAR ON INVASIVE SPECIES, few victories have been more inspiring than the recent successful eradication of goats from Santiago Island, one of the crown jewels of the Galapagos island chain, made famous by its resident finches (which are actually tanagers) and the writings of Charles Darwin. A protected area because of its unique wildlife, Santiago's native ecosystem had nonetheless been dramatically transformed by intentionally introduced species, primarily donkeys, goats and pigs. The combined effects of these large herbivores converted an island that was once covered by trees, cacti and wildflowers into one almost entirely cloaked in grassland.

Getting rid of the goats, which followed successful elimination of the pigs, wasn't easy. The campaign lasted six years, cost US$6.1 million and involved killing close to 80,000 animals. Particularly challenging was the necessary task of getting rid of every last breeding pair. During the project's final phase, researchers shot goats from helicopters; they used radio-collared goats to lead them to the last scattered herds; they released sterilized female goats to lure unsuspecting males; they brought in hunting dogs. More than a third of the project's cost came from the efforts to kill off the final 1,000 animals. But the work paid off. Thousands of hours spent surveying the island in 2006 and 2007 failed to find any sign of goats. In a paper published in 2009, the population was officially declared eradicated.

Currently the saga of Santiago stands as the largest victory ever against an invasive species, in terms of both the area involved and the numbers of animals killed. But it isn't the only one. Indeed, in recent decades conservationists have become increasingly bold in their ambitions. Thanks to growing public support and financial backing, invasive species — particularly rats and other rodents, goats and feral cats — have been exterminated from a list of islands worldwide that now numbers in the hundreds. Because of these and other, similar results, some conservationists are confident that the tide can be turned on invasive species.

But others aren't so sure. While it's true that major victories have been won, it's also true that most of these have involved larger species such as mammals that have invaded contained spaces such as islands, and almost always they've come at a huge financial cost. Such species represent only a tiny fraction of the invasions now spreading around the world. Considering how many aliens have invaded the Mediterranean Sea, San Francisco Bay, the western rangelands of North America and the rain forests of Hawaii, one could see little reason for optimism.

This bleak outlook has led some experts to conclude that humans have steered the planet into a new era. Some have called it "the new Pangaea," in recognition of the degree to which human trafficking of life-forms has turned the world once again into one interconnected habitat. Others are calling it the "Homogecene" — a new epoch marked by increasing homogenization of the natural world. The grim prognosis offered by such visions is a future in which the biodiversity of Earth has been devastated through mass extinctions accompanied by the spread of the same few species into previously distinct environments all over the world. The gloomiest predictions involve complete ecological meltdown. At best they portend a planet that is truly uninspiring — an ugly, impoverished place where nature has been wiped clean of variety and richness, in the same way that world cultural diversity has been leveled by the rise of McDonald's and Starbucks and rap music. It will become, as noted natural history writer David Quammen once put it, a planet of weeds.

But an odd thing is happening. The ecologists and conservationists who set up studies and turned their watchful eyes toward this coming apocalypse have yet to see firm evidence that nature is undergoing the collapse so many have been predicting — at least, not as a direct result of the rise of invasive species. While humans continue to wreak havoc on biodiversity through habitat destruction, overuse of resources and alteration of Earth's chemistry, invasive species as a whole appear to be doing the exact opposite of what conservationists have accused them of doing. Perhaps the biggest shock is that most invasive species don't appear to be having that much of an impact on local biodiversity. Indeed, in habitats around the world researchers are now documenting a surprising *increase* in diversity associated with heavily invaded environments. These results are challenging prevailing notions about the way ecosystems function and the impact of invasive species. Instead of destroying the biosphere, invasive species may be creating a new one from the rubble of our own destruction.

It could be, of course, that it's still too early to tell. But another possibility

— at least, judging from some unexpected observations and an increasingly vocal, small but growing cadre of dissenters — is that the doomsayers have got it all wrong. One place where surprising results have come to light is the Caribbean island of Puerto Rico. A self-governing territory of the United States and one of the world's most densely populated countries, Puerto Rico has a long history of environmental use and abuse. When Columbus landed there in 1493, the island was almost completely covered by tropical rain forest. During the ensuing centuries, gradual clearing of land for timber production, pastureland, agricultural products such as sugarcane, tobacco and coffee, and residential and rural development resulted in almost complete deforestation. By the 1960s trees covered just 5 percent of the island. Pristine forests — forests growing on land that had never been disturbed — had dwindled to nothing more than a few isolated stands hidden away in gullies or clinging to unworkable cliff sides.

Beginning around the mid-20th century, however, social shifts triggered large-scale urbanization in Puerto Rico. The result was relatively sudden and widespread abandonment of pastures, farms and plantations. Over what eventually amounted to more than half the island, nature was again allowed to run wild — or at least, what was left of nature. Centuries of human use had resulted in habitats that were much different from what they had been. Wetlands had been drained; irrigation channels had been dug; rivers were clogged with sediment; soil had been compacted or eroded. The ground lay drained of its fertility and scorched by exposure to the full force of the tropical sun.

Faced with such altered conditions, native species foundered. However, alien species that had been brought to the island over the years thrived. Abandoned pasturelands were overrun by highly invasive species such as African tulip trees and guava trees. Abandoned plantations that had been managed for shade-grown coffee saw widespread proliferation of rose apple, a fast-growing, shade-tolerant invader from Southeast Asia. Land that had been bulldozed was frequently claimed by white siris, an Asian tree that can spread at a phenomenal rate — up to 17 acres (7 ha) a year. From the standpoint of biodiversity, it was not a promising picture. Puerto Rico had entered the Homogecene — or so it seemed.

Since the 1980s, plant ecologists, led by Ariel Lugo at Puerto Rico's International Institute of Tropical Forestry, have conducted systematic surveys of the forests that have been maturing in different microclimates around the island. They've found a surprisingly rapid rate of regeneration and an equally surprising degree of plant diversity. In many places tulip trees are still the dominant species among older trees, but growing beneath the canopy is a lush understory that includes not just invasive species but also thriving populations of many natives. Island-wide, the list of most common tree species now consists of eight natives and five nonnatives. What's more, the balance of power in many of these forests appears to be shifting with time. The shady understory of the maturing forest provides native saplings with an advantage over invaders such as young tulip trees, which thrive best in full sun.

These results run contrary to the notion that ecosystems are closed systems with a limited number of niches, and to the idea that adding nonnative species necessarily entails subtraction of native ones. In Puerto Rico's original forests, researchers have recorded up to 52 different tree species in specific types of habitat. In the new forests, surveys have recorded as many as 74 different species in a single habitat. If you consider all the types of forested areas combined, 750 different tree species are now growing wild on the island, compared to the 547 estimated to be present when Columbus arrived — an increase of 20 percent. Biodiversity hasn't withered under the pressure of sustained invasion. It has expanded.

What's more, these emerging forests appear to be in good ecological health. The density, height and width of the trees suggest high levels of primary productivity — the all-important production of organic compounds through photosynthesis that in most ecosystems forms the foundation of the food chain. The high concentrations of nitrogen and phosphorus found in leaves suggest that nutrients are being efficiently recycled. These and other findings suggest that Puerto Rico is witnessing the birth of new forest ecosystems: mixtures of native and nonnative species that have sorted themselves into communities that are just as much *forest* as any that existed before.

One could argue that a similar situation has occurred in the aquatic habitats of the Baltic Sea, only on a larger time scale. The disturbances in this case were caused not just by humans but also by nature, beginning around 11,000 years ago, when the sea emerged as a freshwater lake at the end of the last ice age. This freshwater period lasted for about a thousand years, at which time rising sea levels connected the basin to the ocean, turning it into a saltwater sea. After another freshwater period that lasted from 8,000 to 7,000 years ago, the sea passed through a lengthy period with moderate salt levels before finally reaching its current brackish state around 3,000 years ago. During each of these shifts it is believed that local ecosystems underwent a complete overhaul: organisms unable to adapt to changing salt levels died off and new ones from connecting waterways invaded. The past 3,000 years haven't been much of a picnic either — the brackish waters are too salty for some freshwater species and not salty enough for most ocean life-forms.

More recently the Baltic's harsh environment became even harsher because of the impacts of humans, particularly eutrophication caused by municipal sewage and agricultural runoff. Since the early 1900s the amount of nitrogen being dumped into the sea has quadrupled, while the amount of phosphorus has increased eightfold. In this case the effects of nutrient overloading — which helped decrease water clarity by as much as 16 feet (5 m) during the 20th century — are exacerbated by the semi-enclosed sea's poor circulation and water exchange with external sources. Periodically the deeper parts of the sea experience buildups of hydrogen sulfide and depletion of oxygen that extinguish all life from the lower depths, causing the Baltic seafloor to be described as "Europe's largest desert."

Besides eutrophication, a likely even bigger threat comes from slowly rusting barrels full of poisons that litter the Baltic seabed. The sea's international waters were long used as a dumping ground for whatever waste the surrounding nations couldn't deal with. The list includes everything from Hitler's remaining stockpile of chemical weapons (some 35,000 tons/31,800 tonnes of mustard gas and other deadly substances dumped at the end of the war) to the 23,000 barrels full of mercury estimated to lie off the Swedish coast alone, a legacy of the 1950s and '60s, when paper mills and other industries routinely (and legally) used the sea as a convenient toxic waste dump.

Despite this backdrop of instability — or perhaps in part because of it — the Baltic has been overrun by invasive species. These include the soft-shelled clam *Mya arenaria* (thought to have been brought from North America by the Vikings during the 13th century), zebra mussels (which arrived in southeastern coastal lagoons in the early 1800s), Chinese mitten crabs, New Zealand mud snails, barnacles and razor clams from North America, more than 30 different species of tiny, shrimp-like amphipods (from large-scale translocation projects carried out by the former Soviet Union in the 1950s to stimulate fish production), marine worms, seaweed and various types of plankton. This list also includes ecologically associated species such as Canada geese, American minks and muskrats, three invaders with partial or indirect links to the Baltic's aquatic ecology. All told, more than 120 invasive species have been reported in the waters of the sea, and roughly 80 have been identified as reproducing successfully — numbers that have been growing with increasing speed since the 1950s.

This massive onslaught completely transformed Baltic ecology. By colonizing hard surfaces, invasive barnacles and mussels have created new habitats for small bottom-dwelling invertebrates. By filtering nutrients from the water and depositing them in the substrate, they've boosted certain food chains. Over the years the empty shells that accumulate on the sandy seafloor have created new habitat that has been colonized by small worms, microscopic crustaceans and the larvae of insects, all components of the diets of many fish. The deep burrows created by the bottom-dwelling marine worm *Marenzelleria viridis*, an invader from North America whose populations have exploded throughout the Baltic basin since their arrival in the mid-1980s, have paved the way for deeper colonization of the sediment layer by other life-forms. These include a variety of other invasive species that now live close to, on or beneath the surface of the seafloor.

In shallow lagoon waters in the southern Baltic, dense swarms of free-swimming invasive mysids and amphipods — miniature shrimp-like creatures that feed on zooplankton and bottom-dwelling invertebrates — provide a vital link in the food chain, gobbling up the rich resources of the substrate and serving as an energy source for fish and other larger forms of life. Since joining the northern Baltic plankton community in

1992, the predatory water flea *Cercopagis pengoi* has become preferred late-summer prey for herring in the Gulf of Finland. Similarly, at least one report indicates that large numbers of round gobies — small invasive fish from the brackish seas of central Asia — have become a favorite food of the larger predatory fish in the southern Baltic. In some places researchers are beginning to see food chains based entirely on nonnative species. In the estuaries of Poland, for example, North American crayfish now feed on invasive plant-like hydroids that feast on the suspended larvae of invasive zebra mussels. Nearby, young invasive North American mud crabs survive on the larvae of invasive barnacles and the aforementioned hydroids before switching to zebra mussels as adults.

Alien species have extended the reach of Baltic ecosystems by invading niches or habitats not previously occupied by any species, native or otherwise. The New Zealand mud snail (*Potamopyrgus antipodarum*) has moved into previously unoccupied sandy areas, where it is able to extract nutrients from bottom sediments. Zebra mussels have taken up residence in microhabitats, such as the deeply sheltered reaches of river-fed lagoons, where the salinity is too low for survival of native mollusks. The Baltic territory that has been overrun by invasive Chinese mitten crabs includes habitats that native crustaceans previously couldn't tolerate.

It is true that certain places — the Baltic Sea and the Great Lakes, for example — now share more of the same life-forms as a result of the spread of invasive species. One could therefore state that homogenization has occurred. But it would be much tougher to argue that the process has impoverished local ecosystems. Indeed, a better case can be made for the exact opposite's being true. In the Baltic's harsh young environments, the most successful alien species have brought life to a broader expanse of physical habitat. They've created new habitats for other species, they've filled valuable links in the food chain and facilitated the flow of energy throughout the ecosystem, and they've broadened the amount of food that's now available to fish living in a variety of habitats.

That's not to say there have been no negative impacts on native species. There's evidence that aliens have reduced the numbers of some species in certain locations — as one would expect of fluctuations between predators and prey in any ecosystem — and they have also eliminated local populations from others. To date, however, no native species have been driven out entirely as a result of the foreign onslaught. Like the forests of Puerto Rico, the aquatic ecosystems of the Baltic appear to be undergoing an ecological renaissance, one in which ecological complexity, proliferation of life in new physical habitats, and overall biodiversity have all been bolstered.

Such resilience is becoming more evident all around the world. An estimated 250 alien species have invaded the various microhabitats found in San Francisco Bay, but scientists have yet to record the loss of a single native as a direct result of this invasion. In New Zealand, invasive mammals have contributed to the extinction of dozens of its unique native birds (mainly because they were flightless, having evolved a terrestrial lifestyle in the absence of predators), but overall its biodiversity has blossomed. While three of New Zealand's roughly 2,000 native plant species have been lost, the list of plants growing on their own in the wild now includes more than 4,000 names, thanks to the spread of more than 2,000 aliens.

The effect has occurred in cities too. A study done in Berlin in 1990 found that 839 native species were complemented by nearly 600 additional nonnative species. And while we're on the subject of nature's robustness in unusual places, consider Toronto's Tommy Thompson Park (also known as the Leslie Street spit), which extends into Lake Ontario. Born in the late 1950s from twisted rebar, slabs of concrete, old telephone poles and construction debris — a post-apocalyptic landscape if ever there was one — the 3-mile-long (5 km) landfill site has become what many people regard as a true wilderness site. Tall cottonwood forests, meadows, wetlands and mudflats are used by a list of wildlife that includes 290 species of birds, 19 mammals, 7 reptiles and amphibians, and a wide variety of fish, insects and plants — a remarkable testimony to the powers of invasion.

ZEBRA MUSSEL

Dreissena polymorpha

FRESHWATER MUSSELS are widely regarded as playing an important role in maintenance of our well-being. As filter feeders they essentially purify our lakes, rivers and streams. But they're also incredibly sensitive. If something goes afoul in their aquatic environment, mussels are often the first species to suffer. This causes something of a dilemma. Humans are naturally drawn to fresh water because we depend on this resource in so many ways: we drink it; we water our livestock and crops with it; we use it for transportation, play in it, harvest from it. And we cool our superheated industrial processes with it and build our cities, towns and vacation homes beside it. The dilemma is that we also use it as a convenient dump for our waste — sometimes treated, sometimes not.

Unfortunately, human needs have usually been met without much concern for the health of the ecosystems in these environments. Indeed, thanks to a long list of major impacts — construction of dams, levees, canals and diversions; increased sedimentation due to erosion of surrounding land; dredging; draining of marshes and other wetlands; dumping of excess nutrients; pollution; overharvesting of fish and other resources — humans have pretty much turned aquatic environments upside down. And, not surprisingly, native mussels have been among the hardest hit. Just about

◀ Unlike other types of mussels, zebra mussels are able to attach themselves to most man-made materials. Since they're able to flourish in polluted waters, zebra mussels have a competitive edge over native species, which typically prefer more natural environments.

everywhere humans have settled in large numbers, these delicate but important organisms have been in steep decline. A recent examination of archeological records found evidence that the decline in North American freshwater mussels dates back to when native people first began clearing land to cultivate plants. For at least 5,000 years, in other words, humans have been making life difficult for mussels.

But now the mussels are getting their revenge. From out of the brackish remnants of an ancient central Asian sea — a cruel aquatic environment with a harsh past that survives today as the Caspian and Black seas — comes *Dreissena*, a genus of freshwater super-mussels unlike any other. Its most infamous member is well-known: *D. polymorpha*, the dreaded zebra mussel. Lesser known but equally formidable is *D. bugensis*, the quagga mussel. Both are spreading along the freshwater veins of entire continents, unperturbed by all but the worst of human disturbance. In some places — including lakes and rivers that had been given up for dead — these invaders are thriving in astonishing densities. So great has been their success that scientists now see their presence not as a sign of environmental health but as a ticking time bomb destined to unleash ecological doom.

That's quite an accomplishment, considering what we're dealing with. Zebra mussels and their relations don't live very long, they are immobile and they're surprisingly small — usually no bigger than the end of your thumb. As far as shellfish go, not very imposing, but their impact has been staggering. Able to attach themselves to virtually any hard surface, they've become an enormous nuisance throughout much of eastern North America. Zebra mussels will colonize every available inch of a submerged pipe, then grow on top of one another until the insides of the conduit resemble the effects of an all-cheeseburger diet on human arteries. Navigation buoys have been known to sink under their accumulated weight. Cars pulled from the bottom of lakes look like they've been tarred with sharp black rubble.

In dealing with the menace, both private enterprises and public agencies have had to open their purse strings. In one 1995 survey, 393 businesses, government agencies and public infrastructure operations in the United States reported spending just over $69 million in zebra mussel-related expenses in the first six years of the invasion alone. And that's small change compared to the stresses on those concerned with the health of freshwater ecosystems. In many places they've displaced native mussels, and they've essentially rebooted local ecosystems, particularly in and around the Great Lakes. It's still not clear what the outcome of these massive changes will be. For one thing, there are signs that many aquatic habitats are now more "native" today than they were before the invasive mussels arrived. On the other hand, ecologists continue to view the zebra mussel invasion as an environmental catastrophe and a potential threat to the integrity of industrial North America's freshwater ecosystems.

The spread of *Dreissena* mussels actually began more than two centuries ago. In the late 1700s zebra mussels were recorded for the first time outside their native home in Hungary's Danube River. In the early 19th century the invaders began turning up in other places, beginning with southwest England in 1824 and, soon afterwards, cities in northern Europe. The following decades saw the mussels spread along river systems throughout England, Germany and France; by the end of the century they had claimed much of central and western Europe, including Great Britain. During the 20th century the invasion continued to spread, moving into Scandinavia, Ireland and countries on the Mediterranean, and they've recently reached eastward into the waterways of the former Soviet Union.

The first sightings of zebra mussels in North America didn't come until 1988, first in Lake St. Clair and then a few months later in Lake Erie, off Ohio. From the size of the shells, scientists guessed they had been there for a couple of years, but the question of when they arrived was quickly replaced by worry over where they were headed. By 1989 zebra mussels were all around Lake Erie, with colonies of up to 3,700 individuals per square foot (40,000 per sq m) in the nutrient-rich waters of the lake's western basin. The following year they were seen throughout Lake Ontario and in the upper reaches of the St. Lawrence River, as well as all of Lake St. Clair, parts of the other Great Lakes, and the Erie and Welland canals. The mollusks took only two more years to spread through the surrounding river systems and conquer a good portion of eastern North America: east to New York City, south to New Orleans, west into the Ohio and Tennessee river systems, and north to the Ottawa River. By the mid-1990s zebra mussels had begun colonizing smaller inland lakes in Ontario, Michigan and New York State. Although the rate of spread has slowed since its early stages, it hasn't stopped. As of 2010, zebra mussels have been seen in 27 states and 2 provinces.

The less famous quagga mussel has proven to be an even more aggressive invader. Similarly striped but slighter larger and paler, quagga mussels have completely replaced zebra mussels in certain areas, particularly in Lake Erie and other parts of the lower Great Lakes. They've also colonized lake bottoms in deeper water; one study found quagga mussels in Lake Ontario at depths of more than 425 feet (130 m) and as far down as 540 feet (165 m) in Lake Michigan.

Several factors have contributed to this dramatic invasion by alien species. One has to do with the continual expansion of global shipping routes. The construction of canals giving ocean-going vessels access to ports along inland lakes and rivers opened new passageways for roaming aquatic organisms, which explains how the invasions spread through Europe and North America. The more challenging leap between continents, which involves not just distance but a transition between saltwater and fresh water, is thought to have been made possible by changes in the shipping industry: beginning in the 20th century, empty hulls were filled with water for ballast instead of soil. The practice, in which water is added in one port and released in another, is now widely regarded as the primary facilitator of intercontinental invasions by aquatic species, and the only feasible explanation for how zebra mussels were able to hop from the Caspian Sea to Lake St. Clair.

Disturbed environments seem to be most vulnerable to invasion, and the North American territory taken over by foreign mussels has to be among the most disturbed habitats anywhere. Consider Lake Erie. In the 1960s it was on the verge of environmental collapse, one of the principal threats being eutrophication — the process of deoxygenation that occurs when water bodies enriched by fertilizer runoff and sewage become overgrown with plankton. The ongoing death and decay of the plankton causes depletion of oxygen in the water, which leads to conditions that fish and other life-forms find increasingly intolerable. Add to this all the pollutants pouring in from the continent's industrial heartland and the result was literally the destruction of life. A dramatic example was the disappearance of mayflies. As larvae these insects live among the sediments at the bottom of lakes and rivers, where they serve as a valuable food source for many species of fish. At one point they were so common that summer swarms of adult mayflies were potential traffic hazards. But then in the early 1950s — possibly because of eutrophication — the insects disappeared almost overnight.

In addition to the insults of eutrophication, acid rain, shore erosion and chemical pollution, several other human factors helped transform eastern North American freshwater ecosystems from their native state. The draining of wetlands not only eliminated a biologically rich habitat but also removed an important water purification system and a natural form of flood control. (Before it was drained in the late 19th century, the 4,800-square-mile [12,400 sq km] Great Black Swamp had buffered Lake Erie's southwestern shore for nearly 10,000 years.) Particularly hard hit were native fish populations in the Great Lakes. Changes to surrounding tributaries and near-shore shallows destroyed habitat that many species depended on for spawning and egg laying. Commercial fisheries, meanwhile, dealt a further blow by overharvesting, heavily depleting the populations of lake herring, lake trout, sauger, lake whitefish and the once annoyingly common lake sturgeon. Blue pike and most of the native herring-like ciscoes were cleaned out completely, while some species — lake trout in Lake Erie, lake herring and native Atlantic salmon in Lake Ontario — were extirpated locally.

Finally, nonnative species helped transform what was left of native ecosystems long before anyone had even heard of zebra mussels. Some of these nonnative species, like brown trout and common carp, were deliberate introductions made during the 19th century. Others were invaders that took advantage of canals. The sea lamprey, a blood-sucking parasite that likely arrived in the early 20th century, nearly finished off the last of the lake trout in much of the Great Lakes. The alewife, a freshwater-tolerant herring from the St. Lawrence River, arrived in the Great Lakes to find a predator-free nirvana. Tougher than the struggling native baitfish, they quickly became the dominant plankton-feeding fish throughout the Great

Lakes. They were completely out of control by the mid-1900s, careering through population explosions and collapses that would leave beaches buried beneath a reeking layer of rotting fish bodies.

Partly to balance what was becoming an increasingly unstable food web and partly to re-engineer an ecosystem that had once teemed with food and income opportunities, humans altered the natural state even further with deliberate species introductions — principally salmon from the Pacific, but also rainbow smelt and white perch. Given nature's usual lack of cooperation, the new and improved Great Lakes ecosystem became dependent on annual stocking of fish bred in hatcheries and then trucked into the wild.

Meanwhile the native mussels were taking a beating. By far the most diverse collection of freshwater mussels found anywhere on Earth — nearly 300 species, most of them from the eastern United States, compared with just 12 in all of Europe — they were an important food source for native North Americans and a useful raw material for making buttons and cultured pearls. It was the disturbances to their habitat that really put them on the ropes. Many of these mussels were highly specialized. Species living in the Ohio River, for example, had evolved in rapidly flowing water that for much of the year was only a few inches deep. After dams transformed a river that at times could be walked across to one that was slow-moving and deep year-round, many of these native mussels could survive only in the few feeder creeks that had maintained their natural characteristics.

An even more dramatic example of how native species depend on their habitat can be found in the life cycle of North American mussels. As larvae, before settling on the bottom and developing into adult shellfish, they live as parasites inside the gills of fish. For life-forms at the mercy of the currents, hitchhiking aboard a highly mobile host is a brilliant way to settle new territory upstream — so brilliant, in fact, that evolution drove some of the mussels into dependency on a single host species. The drawback, of course, is decreased ability to weather environmental change. Ebony shell mussels living in the upper Mississippi Basin exemplify the fate that befell many. Thought to live primarily in the gills of the skipjack herring, this once common mussel went into a steep decline after dams interfered with the spawning runs of its host.

Taken together, the impact of humans has had devastating effects. Habitat alteration is thought to have caused the loss of between 30 and 60 percent of native freshwater mussels from many U.S. rivers. Studies have shown that broad shifts occurred among native populations as the land around the Great Lakes was being converted from forests to farms, with declines among some once-common species and range expansion among certain others. Not surprisingly, the opportunistic natives were generally mussels that could live in silt-laden substrates and withstand a greater degree of pollution. Other research has further revealed the sorry state of native mussels. In western Lake Erie their density declined from ten per square meter in the 1961 to six in 1972 and four in 1982. In 1961, 94 percent of the survey sites had native mussels; by 1982 the figure was down to 35 percent. Native mussels were clearly in a dire state, and the door to invasion was open.

Several factors explain why *Dreissena* mussels have been so successful at invading habitats that many other mussels find distasteful. Their larvae don't rely on fish, for one thing. This means they're more dependent on currents for dispersal, but also that their survival isn't tied to the well-being of another species. Having free-swimming larvae is likely also a major reason why zebra mussels are able to survive being transported long distances in ballast water. Another big difference between zebra mussels and native mussels is surface preference. While native mussels, much like clams, burrow into the soft substrate, zebra mussels are more like true ocean mussels in that they use sticky elastic threads, known as byssus fibers, to attach to just about any hard surface. This includes not only rocks but also the shells of crayfish and of other bivalves, and just about any material that's been invented by humans — nylon, fiberglass, wood, plastic, metal, concrete. Although larvae transported in ballast water are believed to be the primary mode of spread over long distances, adults attached to boat anchors and hulls have been identified as a source of zebra mussels in lakes at a distance from major river systems.

There's no doubt more to it than just sticky fibers. Both zebra and quagga mussels are extremely hardy compared to other mussels. They can survive in water that's slightly salty and they can tolerate wide ranges of water quality, depth, light intensity and temperature. This hardiness may be a result of having evolved in a rough neighborhood. Recent research has suggested that one reason why so many invasive species in the Great Lakes come from the species-poor Black and Caspian sea basins — some 70 percent of the roughly 200 species — may be the turbulent environmental history of that part of the Old World.

Once part of a vast inland lake around 30 million years ago, the region has experienced frequent major upheavals because fluctuating water levels have repeatedly opened and closed connections to surrounding seas. This has resulted in both short- and long-term fluctuations in water temperature, oxygen levels and — even more important — salt content. These periodic connections with other seas also permitted frequent invasion by outside species. Taken together, the harsh conditions may have limited what types of plants and animals were able to survive and evolve — hence the region's lack of diversity. Biological specializations such as those found among native North American mussels would have been punished by extinction. The survivors would have been hardy generalists capable of withstanding a turbulent environment. To these life-forms, the battered freshwater habitats of Europe and North America must have seemed a lot like home.

Zebra mussels have certainly thrived in North American waters. Shortly after their initial discovery in Lake St. Clair, scientists conducted a survey to gauge the extent of the invasion. In some places they counted as many as 18 mussels per square foot (200 per sq m); when they went back a year later the maximum density had reached 420 individuals per square foot (4,500 per sq m). In western Lake Erie the numbers were even more astounding. After recording a maximum density of 9 mussels per square foot (100 per sq m) early in 1989, researchers returned a few months later to discover invasion rates that were hard to grasp: along the water-intake pipe of one power plant they found approximately 70,000 mussels per square foot (750,000 per sq m).

As these early colonizers grew and ate their way through the original abundant supply of food, the densities (not surprisingly) fell. Even so, throughout much of the lower Great Lakes, most hard surfaces ended up completely encrusted with jagged, tightly packed shells measuring several

This water pipe has been clogged by a colony of mussels. Zebra mussels can easily attach themselves to any hard surface and routinely cause damage to submerged objects.

inches deep in some places. And it didn't end there. By the mid-1990s researchers discovered that the microscopic larvae are capable of colonizing soft, sandy substrates by affixing to a single grain of sand and then using their sticky threads to cement together larger clumps of sediment. Once one mussel has begun this process, others are able to glom on until what emerges is a continually expanding mussel reef. When sonar devices were used to scan the floor of Lake Erie between New York and Pennsylvania in 1995, they found roughly 770 square miles (2,000 sq km) of the lake's soft sediment covered by zebra mussels.

While businesses and lakeside residents struggle with clogged intake pipes and encrusted buoys, biologists have been busy cataloguing the invasion's ecological impacts. To many of North America's beleaguered mussels, the arrival of foreign mussels represented the final nail in the coffin — and an ugly one at that. During the invasion's early years scientists would come across native mussels completely encrusted with thousands of tiny zebra mussels. A year or two later, the native mussels would be gone.

It's thought that the native mussels simply couldn't function beneath the weight of the onslaught; they were, for all intents and purposes, mobbed to death. The result was almost total elimination of native mussels from Lake St. Clair, the Detroit River, the upper St. Lawrence River and western Lake Erie.

Other impacts, meanwhile, have been much broader. As filter feeders, mussels obtain all of their food — mainly bacteria, single-celled algae and various other forms of plankton — by siphoning water through bronchial chambers and collecting most of the microscopic particles. While the edible particles are ingested, the inedible ones are coated with mucus and excreted as "pseudofeces." Zebra mussels, it turns out, are extremely efficient at sucking in water and removing its contents: each adult is capable of filtering 1 quart (roughly 1 L) of water each day. During the mid-1990s, scientists studying Lake Erie calculated that there were enough mussels to siphon the lake's entire volume of water once a week.

One effect of this process has been an increase in water clarity due to removal of suspended particles, which has been recorded almost everywhere that zebra mussels flourish. By removing large amounts of suspended plankton from the water and depositing similarly large amounts of feces and pseudofeces on the lake floor, zebra mussels have effectively redistributed nutrient wealth from one aquatic habitat zone to another. Greater water clarity also enhances the penetration of sunlight, which speeds photosynthesis and increases the depth at which plants are able to grow. And it makes it easier for open-water predators to hunt their prey.

Although many of these changes could be seen as a beneficial continuation of efforts begun in the early 1970s to clean up the lake and stave off eutrophication, researchers were nonetheless worried. They feared that by transferring nutrients from open water to the substrate, zebra mussels would starve smaller plankton-loving baitfish. This would in turn trigger a collapse in populations of larger, commercially important predators, particularly salmon. The prediction was disaster for Great Lakes ecosystems and collapse of the fisheries, causing the loss of hundreds of millions of dollars each year.

On some fronts the concerns were justified. In addition to recording large declines among native mussels, researchers have noted dramatic widespread declines among tiny crustaceans known as diporeia, which feed on phytoplankton and serve as an important food source for mid-level food-web fish such as alewives, bloaters, smelt and sculpins. In Lakes Huron and Michigan, dramatic declines among alewife populations — blamed partly on zebra mussels but also on bad weather and overstocking of hatchery-bred salmon and trout — have resulted in tougher times for salmon fishermen and those operating fishing tours. There is also evidence that lake whitefish in Lakes Michigan and Ontario, which feed heavily on diporeia, are also suffering.

At the same time, there are positive indications. In the lower Great Lakes native mussels have been able to withstand the invaders in some habitats, particularly in certain shallow-water areas and natural wetlands. In places such as the Hudson River they have even managed a modest recovery, leading to hopes that natives and invaders will be able to at least coexist as they have managed to do in many parts of Europe. As for fish populations, the early predictions of catastrophic losses have so far failed to materialize. Indeed, many species that live near the bottom, for example, appear to be thriving. One study done in the Hudson River found that the total biomass of fish in the near-shore zones had risen by almost 100 percent, compared to a 28 percent decline among fish living in the open water.

In Lake Huron, declines in nonnative alewives and salmon have been countered by apparent increases in native walleye and lake trout. In Lake Erie the lake whitefish population, which had been severely depleted since the 1960s, rebounded sharply: annual commercial harvests expanded by four to eight times throughout the first 15 years that followed the invasion. Once among the top commercial species in the Great Lakes, this species returned in 2005 to spawn in the Detroit River after an absence of almost 90 years. Meanwhile, dramatic improvements to the smallmouth bass of Lake Erie — in both size and number — have turned this once-polluted backwater into the most productive smallmouth bass fishery in the world.

Lake Ontario is seeing some positive signs as well. In 1998 a deepwater sculpin was pulled from the U.S. side of the lake for the first time in half a century, and since then the small but growing number of the fish being caught each year during bottom trawls has led to hopes that this formerly widespread native is also on the rebound. In the summer of 2009 scientists discovered young Atlantic salmon in New York's Salmon River. This was the first sign of successful reproduction in over a century for the species in what had been one of its principal spawning grounds; recent spawning activity in other tributaries, such as the Credit River, may also signal a historic turnabout for what was once the largest freshwater salmon population in the world.

In addition to better water quality and improvement in some Great Lakes fish populations, other changes during the past 20 years include a resurgence of bottom-growing plants (including formerly rare native species), an increase among ducks and other birds linked with freshwater food chains, and a blossoming of deepwater life in general. In the early 1990s, much to most people's surprise and the delight of ecologists, mayflies returned to Lake Erie's western basin after a 40-year absence.

Determining the role zebra mussels played in these changes is difficult, particularly because of factors taking place at the same time, including efforts to limit nutrient flow into the Great Lakes and local conservation projects involving habitat restoration and restocking. But some scientists are now viewing zebra mussels as a potential blessing in disguise. This is particularly true in Europe, where fear of the invaders appears to have dulled with time. In Holland, for example, eutrophic lakes have been successfully purified following the intentional introduction of zebra mussels. Elsewhere in Europe scientists have shown that zebra mussels can significantly reduce levels of pollutants in water being discharged from chemical treatment plants.

Even in North America there is recognition of the fact that 20 years of zebra mussels have coincided not with the collapse of native ecosystems but something of a revival. As a recent U.S. Environmental Protection Agency report put it, "Recent data indicates that the structure of Lake Ontario's offshore fish community is changing in response to improved environmental conditions, and that the direction of that change is toward a fish community that more closely resembles that which existed historically."

Also worth noting is that recent fish trends in places such as Lake Huron — declines among some nonnatives and increases among natives — suggest a step toward, not away from, a more natural state.

That said, most scientists and many environmentalists remain on storm watch, convinced that the changes unleashed by invasive mussels will sooner or later lead to problems. Some think that trouble is already here. Invasive mussels are now viewed as the prime suspects in a botulism epidemic that killed tens of thousands of birds throughout the Great Lakes during the past decade. It has been hypothesized that mussels concentrate toxins during filter feeding and then pass them on to mussel-eating birds. However, by 2009 there was still little evidence to support this idea and some evidence against it, including botulism outbreaks in pre-invasion times and the presence on the list of victims from several bird species that never eat mussels.

There are also fears that zebra mussels are now threatening to turn freshwater lakes across eastern North America into poisonous soups by intentionally avoiding certain cyanobacteria (blue-green algae) known as *Microcystis aeruginosa.* These free-floating microbes, which are known to sometimes produce a nasty toxin that's strong enough to sicken a human or even kill a dog, are thought to be blooming more frequently, and there is mounting evidence that such explosions are in some way tied to the presence of zebra mussels. But the case isn't closed quite yet. In Holland researchers have shown that zebra mussels are quite content to dine on *M. aeruginosa,* and there's even a hint that these filter-feeders are somehow capable of safely removing the deadly toxins from the food web.

Finally, and most ominously, invasive mussels are a suspect in one of the biggest mysteries around efforts to rehabilitate the Great Lakes: the return of a "dead zone" in the central portion of Lake Erie. While other parts of the lake are teeming with life, the central basin has periodically experienced low levels of oxygen since the early 1990s, and researchers have hypothesized that it may be linked to invasive mussels' having altered the lake's nutrient cycles. But despite reports that dramatically superimpose invasive mussel colonies on satellite photos depicting the lake's (clearly visible) mid-region problems, researchers remain baffled by what's going on.

Who knows? One day the highly industrious foreign mussels may indeed destroy our vital freshwater habitats in the ways that scientists have been predicting. The changes they have unleashed already, after all, have been significant. For the time being, however, the invaders have proven to be less a villain than a shining example of how we tend to view changes that are out of control in a negative light. Like so many other invasive species, zebra and quagga mussels will remain guilty until proven innocent.

JELLYFISH

Phylum Cnidaria and Phylum Ctenophora

ANY WAY YOU LOOK AT IT, the world's oceans are taking a beating. In the past few decades the accelerated harvesting of marine resources has resulted in the collapse of local commercial fisheries. The dramatic rise in the amount of fertilizer and other pollutants being dumped into the water since the end of the Second World War has resulted in the creation of near-shore "dead zones." Increased carbon inputs into the environment have even begun to alter seawater chemistry. And yet, despite all this, there have been no real signs of serious damage; ocean life just seems to carry on.

Now, however, here come the jellyfish. In the Sea of Japan they show up almost every summer — giant Nomura's jellyfish that can grow as big as sumo wrestlers — gathering in such large numbers that trying to drag a fishing net through the water becomes impossible. And that's just the beginning. In 2005 a nuclear power plant in Sweden was forced to shut down when unusually large numbers of jellyfish were sucked into its cooling tanks. In recent years unprecedented numbers of mauve stingers have forced beach closures along the Mediterranean and caused thousands of bathers to seek treatment for stings. In 2007 billions of the same species — reportedly covering 10 square miles (26 sq km) and 35 feet (11 m) deep — drifted into Northern Ireland's only salmon farm, killing an estimated 100,000 young salmon and causing some US$1.6 million in damages. In October 2008, moon jellyfish shut down the Diablo Canyon Nuclear Power Plant in California after clogging its water intake system. Other regions that have reported unusually high numbers of jellyfish in recent years include

Tasmania, China, Korea, Namibia, Hawaii, Alaska, the eastern seaboard of the United States and the northern part of the Gulf of Mexico. There have also been outbreaks in the Adriatic, Black and Caspian seas.

Although scientists still aren't sure exactly what's going on, there is increasing fear that rising jellyfish numbers are a sign that the world's oceans are undergoing a fundamental ecological shift, likely in response to human activity. Some researchers even hypothesize that we're edging dangerously close to a major tipping point — a transition from highly evolved food webs, topped by large predatory fish and sea mammals, to more primitive ecosystems in which the dominant predators are jellyfish, the creatures thought to have ruled the seas some 500 million years ago. What's more, scientists have reason to believe that such a shift would be irreversible.

This scenario is disturbing partly because jellyfish are so primitive. They have no bones, no external skeletons, no blood. They can't see, hear or smell. Their life cycle often consists of several stages, including a polyp state, where they attach themselves to a hard surface and stay in one place almost like plants, and an adult phase, in which they're mobile yet totally at the mercy of the currents.

The term *jellyfish* is a semi-official one, often used to describe groups of organisms that belong to two separate phyla. One group, the cnidarians (pronounced *nid-AIR-ee-ahns*), is represented by species such as the highly venomous box jellyfish and includes most of the true jellyfish familiar to beachgoers. The second are the ctenophores (*TEEN-uh-phores*), also known as comb jellies. They are a highly diverse collection of species that range in size from tiny specks to giants such as Nomura's jellyfish and the lion's mane jellyfish, a coldwater species with tentacles that can grow up to 120 feet (36.5 m) long.

Despite their primitive nature, jellyfish are carnivorous predators. They'll feast on anything that happens to float their way, including plankton, shrimp and crab larvae, fish eggs, small crustaceans, other jellyfish and even small fish. Their inability to chase down prey is compensated for by

This massive school of moon jellyfish was photographed in the Irish Sea. ▶
The jellyfish emerged in response to a plankton bloom, and such
blooms seem to be becoming larger and occurring more frequently.

ingenious specialized weapons. Ctenophores, for example, trap their prey by discharging a glue-like substance. Cnidarians paralyze their victims by using venom-producing stinging cells located along their tentacles. Among some species, such as the box jellyfish that live off the coast of Australia, these toxins are powerful enough to kill humans in only a few minutes.

Scientists have had a tough time trying to discover whether recent increases in jellyfish populations are really worth worrying about. On the one hand, jellyfish are known to proliferate rapidly in response to positive changes in prey abundance or environmental conditions such as water temperature and sunlight. The size of these "blooms" can vary from year to year. On the other hand, these population explosions are occurring in many places on a scale now widely viewed as unprecedented. In the Sea of Japan, for instance, Nomura's jellyfish are known to have drifted in from the south in large numbers three times during the 20th century: in 1920, 1958 and 1995. Beginning in 2002, however, they've turned up every summer but one, and in astonishingly high numbers. In 2005, one of the worst years, up to 500 million Nomura's jellyfish were reported to be drifting into the sea each day.

Several factors have now been identified as possible contributions to the increased success of jellyfish worldwide. One of the leading suspects is the exploitation of fish and other marine resources, something that has intensified in recent decades, partly because of advances in large-scale seafood harvesting and processing techniques. Only a handful of species are thought to prey directly on jellyfish, and most of these predators — including giant sea turtles — are becoming increasingly rare. The main impact of overfishing, though, may stem from the reduction of filter-feeding fish such as sardines and anchovies, which eat the same food as jellyfish. In the southern Atlantic waters off Namibia, where overharvesting has resulted in complete collapse of a once-thriving sardine fishery, unusually large numbers of jellyfish are now a permanent feature of the near-shore marine ecosystem.

At the same time, jellyfish seem to thrive under conditions that are becoming increasingly widespread because of human-associated activities. Although the theory is highly speculative and still under debate, global warming and acidification of the oceans — a result of more carbon dioxide dissolving in the water — may be two such factors. Jellyfish love warm water, for one thing. And at least one study, from the North Sea, has reported finding a connection between greater jellyfish abundance and lower pH levels.

Another environmental change, and one that is more firmly linked to expanding jellyfish populations, is the eutrophication that occurs in near-shore ocean waters close to large human population centers and the mouths of large rivers. A strong correlation exists between blooms of algae and other plankton, caused by excessive nutrients from sewage and fertilizer runoff, and jellyfish eruptions. One example is the waters off the southern U.S. coast. Nutrient-enriched waters from the Mississippi River have created a massive dead zone in the Gulf of Mexico.

While fish and other aquatic life-forms are finding survival increasingly challenging, jellyfish such as moon jellyfish and sea nettles are becoming increasingly numerous. This is not really surprising. A checklist of jellyfish traits — good survival rates during periods of starvation, rapid reproduction, diverse diet, ability to feed in murky water, capacity for surviving under low oxygen conditions typical of dead zones — reveals that these seemingly fragile organisms are actually tough life-forms ideally suited to survive in disturbed environments. According to the National Science Foundation, the main science funding agency in the United States, there are currently 400 known ocean dead zones where almost nothing lives except jellyfish.

Finally, like many other marine invasive species, jellyfish have benefited from the globalization of human trade. Many species, particularly ctenophores, are hardy enough to survive in the ballast water of ships, and jellyfish polyps can attach themselves to their hulls. Because of this, jellyfish species are being inadvertently introduced into new aquatic habitats. One of the best examples involves the warty comb jelly (*Mnemiopsis leidyi*), a ctenophore that likely made its way into the Black Sea for the first time in 1982. Since then this highly adaptable species has been thriving; its original nonnative range has expanded to include not just the Black Sea but also the Caspian, Baltic and North seas. Similarly, over the past 30 years the Australian spotted jellyfish (*Phyllorhiza punctata*) has been introduced from the western Pacific Ocean into the central and eastern Pacific, the Mediterranean, the southwestern Atlantic, the Caribbean and the Gulf of Mexico and along much of the southeastern U.S. seaboard. The species has done particularly well in the Gulf of Mexico, where aerial surveys in 2000 estimated there to be 5 million jellyfish, weighing a total of 44,000 tons (40,000 tonnes), in one 50-square-mile (150 sq km) area.

While it's unclear which factor may be contributing most to the rise of jellyfish, it's probable that what's happening is a result of several factors in combination. The case of the warty comb jelly serves as a good example. Although it was introduced into the Black Sea as early as 1982, it remained a minor player in the sea's ecosystem throughout the 1980s. A significant expansion of the anchovy fishery during that decade eventually caused the fish to enter a steep decline after 1989. After the crash of the anchovy population, plankton levels rose and jellyfish populations exploded.

Slightly smaller than Nomura's jellyfish, Echizen jellyfish grow up to 5 feet (1.5 m) in diameter and are just as much of a menace to fishermen in Japan.

Around this time the Black Sea was also experiencing increasingly severe eutrophication, which appears to have exacerbated the situation. Evidence supporting this idea came following the collapse of the Soviet Union, when a sharp decline in fertilizer use led to noticeable improvement in the quality of water in the Black Sea. Sure enough, this was followed by a reduction in overall numbers of jellyfish.

Interest in getting to the bottom of the mystery is intensifying now that jellyfish are becoming increasing nuisances around the world. The annual onset of Nomura's jellyfish is now a major headache for Japanese fishermen. The water-laden jellyfish, which are impossible to avoid when fishing with nets, weigh down and damage equipment. Their stingers kill and damage fish caught in the same net, reducing their commercial value. Similarly, recent blooms of spotted jellyfish are being blamed for millions of dollars in losses suffered by the Gulf of Mexico shrimp industry. The jellyfish are believed to compete with shrimp for food, and when they bloom they also make it difficult (sometimes even impossible) for shrimp boats to operate.

Fishermen aren't the only ones suffering. Industrial operations that rely on water intake for cooling or other purposes are being increasingly slimed by mobs of jellyfish. Besides causing nuclear power plant shutdowns in at least four countries, jellyfish have recently interfered with the works at diamond mining facilities in Namibia and desalination plants in Iran. They've also clogged the intake pipes on ocean-going vessels. And jellyfish-related problems are plaguing the tourism industry as well. Increasingly blooms are keeping people out of the water in resort areas of the Black and Mediterranean seas, as well as along the northeastern coast of the United States. In Australia there is growing concern that the deadly box jellyfish, which has long kept people out of the water for much of the year in northern Queensland, may be expanding its range further south.

What's really grabbing attention, however, is the possibility that recent trends are part of a major upheaval in the world's ocean ecosystems. One potential problem is that as humans reduce fish stocks, jellyfish proliferate in ways that make natural recovery of those stocks difficult and perhaps even unlikely. Early research conducted on herring in the North Sea demonstrated that jellyfish affect fish populations by competing with them

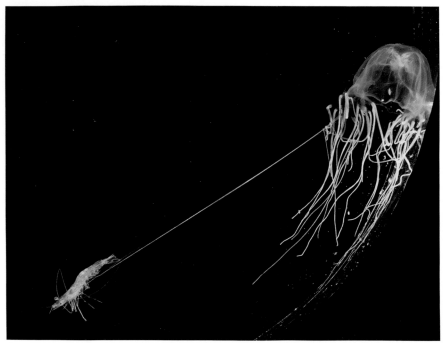

This box jellyfish has snagged a prawn, which died instantly from the jelly's venomous sting. The box jellyfish's toxins are so potent that they can kill a human within minutes.

for food and by preying on their eggs and larvae. This has led to the theory that as fish stocks dwindle they eventually reach a point where there's a role reversal between predator and prey. When there are large numbers of adult fish present, jellyfish can't compete. When the adult fish are removed, as they are by heavy fishing, jellyfish numbers are allowed to increase. Eventually the jelly hordes are eating most of the fish eggs, permanently suppressing their former rivals.

Circumstantial evidence now suggests that this series of events has occurred off the coast of Namibia. In this highly productive marine habitat, large jellyfish are believed to have been present in only insignificant numbers before the 1960s. They began to appear in somewhat larger numbers after this period, which coincided with an El Niño event and initial declines in commercial fish stocks. Since the 1990s an even more noticeable increase in jellyfish numbers has taken place, in which jellyfish appear to have replaced fish as the dominant form of marine life along the Namibian coast. While commercial fish stocks have remained imperiled, despite stricter government controls on commercial fishing imposed in 2006, the jellyfish have taken over. In 2006, researchers from Scotland and South Africa published the results of a study in which they used echo sounders to survey the biomass in nearly 34,000 square nautical miles (88,000 sq km) in the continental shelf zone off the coast of Namibia. The team calculated there to be around 13.5 million tons (12.2 million tonnes) of jellyfish, compared to only 4 million tons (3.6 million tonnes) of fish. The scientists concluded: "Jellyfish play potentially major controlling roles in marine ecosystems and, in this era of apparent jellyfish ascendancy, marine ecosystem managers and modelers cannot afford to ignore them."

On an even more fundamental level, some scientists have interpreted the recent rise in jellyfish as a sign that ocean ecosystems may be undergoing an ecological "regime shift" — basically a wholesale restructuring of the marine food web. The basis for this idea is the theory that the complexity of marine ecosystems can vary depending on the efficiency with which energy is passed along food chains. In some parts of the ocean — because of several factors, including water circulation and nutrient cycling — the dominant life-forms at the base of the food chain are single-celled algae such as diatoms that are too large to be eaten by other single-celled organisms. Instead they support large numbers of copepods, tiny crustaceans that in turn can be eaten by — and form an important part of the diets of — small fish, whales, seabirds and larger crustaceans. Other ocean habitats, however, consist of cyanobacteria and other primary organisms that are small enough to be filtered from the water by other microscopic protozoa. These single-celled predators are preyed upon by slightly larger protozoa such as ciliates, which in turn are eaten by copepods.

According to the theory, the first system supports highly mobile top predators such as whales and large fish, because energy is transferred up the food chain quickly and efficiently. In the second system, however, microbes use up a much greater proportion of the total energy supply. As a result, there's not enough suitable prey to meet the energy demands of plankton-eating fish, whales or birds. The position of top predator in such a system is thus confined to life-forms that possess the rare combination of large size and low energy demands — in other words, jellyfish.

Five hundred million years ago, food chains of the low-energy variety may have been the only ecosystems in the oceans. Back then, Earth was warm and only recently oxygenated. The water environment was filled with cyanobacteria and was even more eutrophic than today. Jellyfish likely ruled the seas. Over the next 200 million years, however, rising oxygen levels and other changes are thought to have helped fuel the evolution of organisms with greater energy demands. The result was the gradual emergence of more complex food webs, consisting of highly mobile species that could easily outcompete jellyfish. The former food-chain kingpins were pushed aside, left to thrive only in places where conditions remained unsuitable for the hyperactive newcomers.

With the recent increases in water temperature, eutrophication, cyano-bacteria and jellyfish, some researchers have argued that the impacts of human activity are now undoing a half-billion years' worth of evolution. This is supported by the fact that nutrients from fertilizer runoff are rich in nitrogen and phosphorus, which promote the growth of cyanobacteria, and poor in silica, which is a key requirement for growth in diatoms. This possibility, coupled with the suppression of depleted fish stocks by established jellyfish populations, recently led a team of scientists to argue that because such large-scale alterations may be exceedingly difficult to repair, the emphasis needs to be on preventing them from occurring in the first place. "It is ironic," the researchers wrote in 2009, "that the same activities that are driving rapid industrialization and technological achievements are threatening to push marine ecosystems way back to the future."

How does one stop a jellyfish tsunami? Believe it or not, much discussion currently focuses on the merits of eating our way out of the mess. Jellyfish have been harvested off the coast of China for centuries. They're purported to have useful medicinal properties for treating arthritis, hypertension and back pain. And when semidried and salted, they make a tasty snack with an appealing, almost crunchy texture; in many places it is now considered a delicacy. Although the existing markets are primarily in Asia, dried jellyfish products now represent a multimillion-dollar business supplied by active fisheries in at least 14 countries, including the United States. According to a report by the United Nations Food and Agriculture Organization, the total global jellyfish harvest went from almost zero to more than 550,000 tons (500,000 tonnes) a year between 1970 and 2000.

Of course, if the nightmare scenarios that scientists fear turn out to be correct, eating jellyfish will do nothing to ease the impact of a world without fish and other marine resources. For this reason nations around the world are being called upon to invest in research that will shed light on jellyfish ecology, including the roles these strange creatures play in marine ecosystems. In the end, such knowledge may be our only hope for avoiding what was recently described as our "gelatinous future."

HOUSE SPARROW

Passer domesticus

IT WOULD BE EASY to dismiss the house sparrow as a parasite. Here is a species that has spread throughout the world in the wake of human development, taking advantage — literally — of the crumbs we've left behind. You can find them in Johannesburg and in Moscow, in Toronto and in Mexico City and many, many places in between. They are a species untroubled by differences in climate or native habitats. All they care about is humans. Where we go, they go, leeching off our success.

But there's more to this little brown bird's invasion than that, as researchers now realize. Having spread so far and so wide over the past 200 years — becoming, it has been said, the most common bird in the world next to the chicken — house sparrows are now experiencing something totally unexpected: they are in the midst of widespread population declines. Do not be mistaken; this is not the end of this highly successful species. On the contrary, *Passer domesticus* remains common in many places and in some areas is continuing to expand, right alongside growing human populations. But in other parts of the world, house sparrow numbers have been in a decades-long freefall.

One result of this surprising turn of events has been sudden concern about a species that until recently was widely regarded as invincible, and nothing more than a pest. Another is the realization that there may be

◀ Sparrows have lived alongside humans for hundreds of years. These sparrows are nesting in the eaves of a building, proving that our home is literally their home.

more to the sparrow's relationship with humans than previously thought. Indeed, as scientists work to unravel the mystery, they're starting to see house sparrows not as simple freeloaders but as components of a wider network of ecological relationships — an urban ecosystem, if you will, that possesses all the complexities and dynamics one would find in pristine old-growth forest or undisturbed prairie grassland.

P. domesticus — also known as the English sparrow — belongs to a group of sparrows with between 15 and 23 members (depending on species classifications). They are seed eaters thought to have originated in Africa roughly 8.5 million years ago. It was initially thought that house sparrows evolved when human agriculture was first being developed in the Middle East, sometime around 10,000 years ago. But another idea — supported by both fossil evidence and DNA analyses — is that they may have originated independently, long before the rise of farming. According to this theory, an early ancestor — already sporting the distinctive black bib that signifies house sparrows and their close kin — spread through northeastern Africa, possibly along the Nile Valley, into the fertile grasslands of the eastern Mediterranean. As these early sparrows spread east and west, the advance and retreat of glaciers during the ice ages may have created isolated sub-populations and the evolution of distinct species. House sparrows would have been one such species, possibly appearing somewhere around 120,000 years ago. There were early humans around at this time, and it is believed that sparrows may have become closely associated with our ancestors while they were still semi-nomadic hunters and gatherers; fossil sparrow bones have been found in caves associated with early Paleolithic people.

Competing theories surround what drove the continued spread of the house sparrow. One possibility is that they advanced through Europe following the expansion of agriculture. Another is that sparrows weren't as interested in our ancestors as they were in our ancestor's horses. Specifically, these opportunistic birds were after the grain that would have been everywhere after humans came to rely on horses — grain being carried, stored or spilt during feeding, and undigested grain in piles of manure. According to this theory, sparrows would have moved northward, following the arrival of domesticated horses, beginning somewhere around

3,500 years ago. Either way, a big change would no doubt have occurred after humans began farming and storing grain. House sparrows are the only temperate-zone sparrows that don't migrate; it has been suggested that the species gave up this habit in order to reap the year-round benefits of life around permanent settlements.

Several house sparrow subspecies emerged once these birds were established throughout Eurasia. One came to occupy most of India and Saudi Arabia; another inhabited the Italian peninsula and several islands in the Mediterranean. But the most successful by far was the one that established itself throughout the rest of Europe, including Scandinavia and the British Isles, and much of Asia all the way to the Pacific. Although it is impossible to define when house sparrows became "invasive," they are known to have undergone a major natural range expansion in historical times, advancing farther into Siberia, Great Britain and northern Europe along with the building of railroads and the spread of human settlements during the past two centuries.

Around this same time, European immigrants in colonies around the world began to feel a longing for the plants and animals they had known at home. According to one historical account, the first successful introduction of house sparrows into North America took place when 100 were brought from England and released in and around Brooklyn, New York, over two springs, beginning in 1852. During the next two decades, at least 20 additional introductions were made, in various locations in the eastern United States as well as Texas, California and southeastern Canada. Meanwhile, house sparrows were also moved from one part of North America to another more than a hundred times. Sometimes the introductions were an attempt to control pests, but often the reason was more whimsical. They had become, in the words of one observer, a fad.

It wasn't long before house sparrows had worn out their welcome. Their tendency to form large colonies, to live around farms and cities, to forage on crops and garbage and to outmuscle native species soon led people to see them as pests. As early as the 1880s some states were already attempting eradication, and by the turn of the century they were almost universally loathed. As the 1903 edition of *The Birds of Ohio* declared, "Without question the most deplorable event in the history of American ornithology was the introduction of the English sparrow. The extinction of the Great Auk, the passing of the Wild Pigeon and the Turkey — sad as these are, they are trifles compared to the wholesale reduction of our smaller birds, which is due to the invasion of that wretched foreigner, the English Sparrow." Ned Dearborn, in an article titled "How to Destroy English Sparrows," which appeared in the U.S. Department of Agriculture's *Farmers' Bulletin* in 1910, was equally uncharitable: "The English sparrow among birds is comparable to the rat among mammals. It is cunning, destructive and filthy."

The house sparrow didn't seem to mind the bad press. By 1915 the species was established in all parts of the continental United States (except for the wild areas of the Rockies) and most of the southern half of Canada from the Maritime provinces to Vancouver Island. By the 1940s its population in the U.S. had reached an estimated 150 million. During the second half of the 20th century the expansion continued, with house sparrows moving north into Canada's Northwest Territories and south through Mexico and Central America. And that was only one part of the invasion. House sparrow introductions were made on other continents, including Australia and South America in the 1860s and 1870s, and in South Africa in the early decades of the 20th century. Most of these introductions took hold and populations began to spread at different rates in different directions. They're now widely distributed on all three continents, and populations have expanded into Indonesia and other parts of Asia outside their original native range. As if all that weren't enough, house sparrows have also invaded major island groups all over the world, in the Caribbean, the South Pacific, the Indian Ocean and even the South Atlantic.

Scientists documenting house sparrow invasions are astonished at the speed with which introduced populations can grow and spread. After the tiny birds arrived in New York they began establishing populations to the north and the south, but their most rapid advance was into the heart of the continent. At its peak during the 1870s and 1880s, the sparrow invasion registered an average advance of 118 miles (190 km) a year. In Mexico they advanced 124 miles (200 km) per year along the Pacific coast, and between 100 and 150 miles (160–240 km) per year in Central America.

Several traits account for this success. One is aggressiveness. At feeding sites, house sparrows will not hesitate to attack any other bird of equal size or smaller. They'll defend their own nests vigorously, and they're notorious for attacking nest sites occupied by other birds. In one well-documented example, house sparrows are believed to have played a major role in the demise of eastern bluebirds in the United States. During monitoring of nest boxes in South Carolina between 1977 and 1982, one researcher came across 20 dead bluebirds that bore evidence of violent attacks: bloodied heads, cracked skulls and various other cuts and bruises. The presence of house sparrows at the scene before and after the attacks, their subsequent takeover of the dead bluebirds' boxes, and the lack of bluebird deaths at sites with no sparrows led the researcher to conclude that 90 percent of the attacks were likely carried out by sparrows. One afternoon a house sparrow was caught red-handed, darting repeatedly into a nest box containing five unattended bluebird nestlings. After an hour, four of the nestlings were dead from head wounds and the fifth died from its injuries the next day. Even more gruesome, sparrows were seen building new nests right on top of the dead bluebirds, incorporating parts of their corpses into their new homes.

In addition to being bullies, house sparrows thrive by being able to eat a wide variety of foods. Scientists sifting through regurgitated stomach contents have come across corn, oats, wheat and other grains; sunflower seeds and various other seeds found in commercial birdseed mixes; a wide range of grass seeds; bits of vegetation including parts of flowers, leaves and blades of grass; a long list of invertebrates, including various beetles, springtails, flies, aphids, ants and spiders; and — no surprise for outdoor café patrons — bread. Sparrows in rural areas seem to be more content with seeds and insects, and seeds are certainly their forte: they have a special bone in their tongue, the preglossale, for husking them. However, in the urban jungle the ability to eat lots of different foods — and to survive on whatever might be available — may play a big role in the house sparrow's success in this highly unusual habitat of concrete and asphalt.

House sparrows are also highly communal. In a 1977 study near a ranch in Alberta, 110 nests were found grouped together in a small stand of spruce trees. The average distance between each nest was only a couple of feet (60 cm), and in some places up to four nests formed a single unit — condo living for sparrows. The ability to tolerate such densities is thought to explain why sparrows can proliferate rapidly once they invade new territory. Living in close quarters enables the birds to utilize all the available nest sites, even if the sites are clustered in one area. And being highly social is also believed to help each member of the colony expand its foraging success, avoid predators and outcompete rival species.

Like many invasive species, house sparrows are also sometimes capable of rapid reproduction. They breed two, three or even four times a year. Each clutch usually ranges somewhere between three and six eggs, although clutches of up to nine eggs have been recorded. Sparrow survival may also be enhanced by a reproduction schedule that is rapid compared to that of many other birds. Eggs are incubated for around 11 days, and fledglings are sometimes ready to fly when they're only 11 days old.

Finally, and perhaps ultimately, there is the house sparrow's ability to successfully exploit human environments. Such environments offer potentially huge rewards for life-forms capable of adopting them as their native habitats. For one thing, they are well endowed in certain resources because humans are always importing food and materials from elsewhere. Second, there is often little competition for this bonanza, because the vast majority of Earth's plants and animals find life in the shadow of humans far too stressful. Bird surveys have shown that as urban environments build up, the total number of species capable of surviving in them goes down, compared to the surrounding native landscape. But, surprisingly, the total number of individual birds goes up. House sparrows, to state the painfully obvious, are one of the main reasons. While they rarely inhabit wilderness habitats of any sort, they seem to be at home in almost every form of human-created landscape. There is even a report from the late 1970s of house sparrows surviving on handouts in a mine shaft in Yorkshire, England, for three years — more than 2,000 feet (640 m) below the ground.

Their ability to exploit human environments seems to be linked to a list of traits that includes a certain fearlessness (like squirrels, wild sparrows will often take crumbs right from human hands and nest happily amid the noisy bustle of the urban environment), adaptability (in diet and where they nest or forage), intelligence (which helps them deal with novel situations presented by this environment) and a lack of wanderlust (which may help them occupy and defend urban territory against outsiders). House sparrows are natural cavity-nesters, so human environments provide plenty of options. They'll nest in cracks between barn boards and under eaves, in rafters and gutters, behind the slats of blinds, attic vents and wood siding, in nest boxes and birdhouses, behind climbing ivy, among the nooks and crannies of commercial signs or atop streetlamps. In India, house sparrows were discovered nesting inside a pair of pants left hanging in a shed. In Kansas, they were found building nests atop oil pumps that were in constant motion. If a pair of sparrows had the opportunity to nest in the backseat of a roller coaster, they probably would. As for eating, they're not at all picky when it comes to birdfeeders, garbage dumps or leftover food, particularly in the urban core, where such resources may provide the only foraging opportunities.

Although they can't equal the brainpower of crows, house sparrows are nowhere near as dumb as pigeons. In the April 1936 issue of *Bird Banding* (now the *Journal of Field Ornithology*), natural historian Jacob Brenckle described how his efforts to design traps for bird-banding studies resulted in new appreciation of sparrows' ability to think their way out of a jam. They were frequently able to escape traps that captured most other songbirds, including one in which escape involved diving into a pan of water and swimming for up to 14 inches (36 cm) to reach an exit. Remarked Brenckle, "While all Sparrows do not 'take to water,' I have seen both young and adults do it. So far no other species of small birds has attempted it." There are also reports of house sparrows triggering automatic door sensors in order to gain entry into places with predictable food resources, such as bus stations and fast-food restaurants.

Finally, house sparrows are the only members of their family from the temperate zone that don't engage in seasonal migrations. Their sedentary nature is thought to have developed when they first began hanging around farms, and it is believed to provide them with several advantages. A major one is that by staying put, house sparrows may be able to monopolize nest sites.

The truly remarkable part of the house sparrow's fondness for human environments is how it seems completely oblivious to the environmental factors that usually define where and how a species can live — the amount of rainfall, daily and annual temperature ranges, the presence or absence of seasons, the amount of time between sunrise and sunset. You can find sparrows off the coast of Norway, north of the Arctic Circle. You can find them in Death Valley, California. You can find them in Ushuaia, Argentina, the most southerly town in the world. You can find them 10,000 feet above sea level in the Rocky Mountains. They're in Caracas, Ottawa, Katmandu and Nairobi. Just about everywhere humans have set foot (short of Antarctica and the moon), house sparrows have felt at home.

Given this track record, it's hard to believe that house sparrows would ever find themselves facing any sort of threat. Expanding human populations have not only continued to alter the face of Earth, they are altering it at an accelerated rate. One would expect house sparrow populations to benefit, but that hasn't been the case. The declines appear to be greatest — or at least most carefully studied — in Great Britain, where watching birds is as popular as drinking tea. Taking advantage of annual bird count statistics, researchers have calculated that house sparrow numbers fell by 50 percent during the final decades of the 20th century, from around 26 million in the early 1970s to about 13 million today. In private gardens, where the problem appears to be worse, populations have been in a steady free fall since around 1983. The situation isn't universally grim — in parts of Wales and Scotland the numbers are rising — but where things are bad, they're very bad. In London's Kensington Gardens, for instance, the sparrow population went from a peak of more than 2,600 in 1925 to just eight birds in 2000, and throughout central London it's now rare to spot a sparrow. Across the U.K. this former pest species is now officially listed as a species of conservation concern.

The problem isn't limited to the British Isles. Studies showing declines in Hamburg, Moscow, Prague and Rotterdam suggest that house sparrows may be having a hard time throughout northern and eastern Europe. Conservationists have also reported a steep drop in sparrow numbers in India and in at least one location in Australia. In North America, several studies have reported major house sparrow declines over the past 30 years, particularly in the northeastern U.S. and Canadian maritime provinces. Statistics from the Audubon Society's annual Christmas Bird Count also show a steep decline, with house sparrow numbers in 2000 less than half of what they were in the 1960s.

What's going on? Some evidence indicates that sparrow numbers in agricultural habitats may have peaked during the early decades of the 20th century. Since then the shift to modern farming practices may have made things more difficult. The internal combustion engine meant the end of reliance on animals for transportation and farm labor, which reduced the availability of grain. Improvements in barn construction and crop storage further reduced the amount of food for sparrows, and possibly nesting sites as well. More recent hygiene and food safety laws that require farmers to keep pests away from stored crops have also been identified as part of the sparrow's problems. Monoculture farming, which limits the diversity of available food, is another culprit, and other factors under suspicion include pollution, pesticides, nestling starvation due to insect control measures, and predation by cats and hawks.

The picture isn't as clear in cities, where declines appear to have intensified only in recent decades. At first researchers wondered if it had something to do with what zoologists call recruitment: the flow of new birds into an area from outside populations. If sparrow populations were declining in the countryside, maybe not enough individuals were left to spill over into towns and cities. But studies have shown that sparrows aren't all that mobile anyway. They rarely migrate during the winter, and when they do disperse, it is usually only a mile or two at most. Researchers have concluded that city and rural populations are largely distinct, which suggests that urban declines may be happening for different reasons.

A newspaper in London was recently moved to offer a reward equivalent to $8,000 to anyone who could solve the mystery. Several theories have also emerged. In 2006 it was reported that birds, including sparrows, were dying in large numbers in England and Wales from trichomoniasis, which is caused by a parasite. Other studies have singled out predation from house cats, pollutants in unleaded gas, and the increased use of insecticides. And then there's electromagnetic radiation. According to a study in Valladolid, Spain, the density of sparrows in that city is inversely related to the strength of emissions from telecommunication towers; the researchers involved suggest that the decline in urban house sparrows may bear some relation to the rise of the cell phone.

While these and other factors may be important, searching for single causes to explain single effects is a traditional way of thinking that may fail to address the reality of the urban landscape. Until recently cities have been viewed as anti-wilderness zones — unnatural places where the laws of ecology no longer apply. Increasingly, however, ecologists have come to appreciate that even though we've paved most of the ground, polluted the air, fouled the water, erected a forest of concrete and glass, and limited vegetation to erratic and fragmented patches, nature is far from dead. Cities still contain microbes, plants and animals (including humans and our pets) and they still interact with each other and with soil and sunlight and moisture and all the other components of an urban environment (including lime from concrete, chemicals from car exhaust and electromagnetism from transmission towers). They are, in short, ecosystems, and the sudden decline of urban house sparrows may be related to changes in those ecosystems.

Modern cities, for example, appear to have become increasingly hostile to sparrow chicks. Between 2001 and 2003 British researchers examined house sparrow reproduction at nest boxes in and around the city of Leicester. The results revealed smaller clutch sizes and reduced chick survival compared to expected averages, suggesting inadequate reproductive output. High mortality in the first week after hatching and following fledging suggested that the baby sparrows were not getting enough to eat. This hints at yet another possible problem: a reduction in the availability of insects and spiders, which make up the bulk of the diet of young birds. This was supported by the study's finding that fledging survival rates were highest in nests located in areas of greater abundance. The researchers also noted a correlation between automobile exhaust pollutants and decreased body size in chicks, which suggests that car traffic may be promoting malnourishment — perhaps due to parent birds being hit by cars, insect habitats being affected by car exhaust, or direct toxic effects from the pollutants.

Other factors might also be contributing to insect declines. For example, house sparrows get their insects and other invertebrate prey from deciduous woody vegetation — trees and grass — rather than the evergreen shrubs and ornamental plants that private gardeners have come to prefer in recent decades. Several other trends, such as conversion of front gardens into parking pads, rejuvenation of abandoned lots, removal of trees and development of sites formerly full of vegetation, may have contributed further to a decline in urban insects and other invertebrates.

But the picture may be even more complex. In addition to evidence of sparrow declines, bird survey results have revealed that the declines have been patchy: more in big cities than in small towns, in some cities but not in others, or only in certain areas of a given city. What's more, this patchiness may relate to human socio-economic status, as wealthy areas have fewer sparrows than poor ones. Two British studies in the early 2000s discovered a strong correlation between the presence of birds and the degree of social deprivation; in wealthy suburbs sparrows were few and far between, but in crowded housing developments the birds were plentiful. In Berlin around the same time, it was found that house sparrows had become rare in the affluent western part of the city while remaining common where development had been curtailed by communist governance in the east.

Researchers have identified several ways in which the fate of the wildlife around us might be tied to how much money we have: how the urban ecosystem can be affected by changes in human society. In the United Kingdom, several changes associated with growing urban wealth may be contributing to the sparrow declines in particular. Increasing demand for off-street private parking (driven by a rise in automobile ownership) has resulted in loss of vegetation from front yards (two-thirds of all the front yards in London are now partially or completely paved over). Neighborhood gentrification tends to attract homeowners with more time and money to invest in their gardens, resulting in changes that may be good for humans but not so beneficial for insects, birds and other wildlife: deciduous native shrubs replaced by evergreen hedges, self-seeding weeds and grass replaced by tidy lawns, and an increase in pesticide use. Government policies aimed at reducing vacant or undeveloped lots — which are usually overrun by weeds and are ideal insect habitat — have likely added to the effect, helping explain the insect shortage that apparently plagues young sparrows in some places.

At the same time, changes to our homes — the types of materials used for construction and repair, how well they're maintained — may be reducing the number of potential nest sites. In Bristol birds often nest on roofs with curved tiles or behind wooden fascia boards, features found mostly in older homes or public housing. Studies from other cities suggest that more frequent repairs and renovations, construction of newer homes and the use of less rot-prone modern materials such as plastic have made it increasingly difficult for sparrows to find a good place to build a nest. Indeed, research from Austria, Spain and throughout the U.K. has demonstrated that breeding sparrows have a distinct fondness for older homes.

Reactions to the decline of sparrows, as one might expect, have been mixed. In Great Britain, where the species is a native, conservationists and government agencies are pushing urban residents to incorporate sparrow-friendly habitat into their landscape designs — by sowing wild grasses and flowers, installing backyard ponds, reducing the use of pesticides, putting up nest boxes, planting trees and hanging out bird feeders. More than a century after English sparrows were exported around the world, partly for insect control, humans are now trying to foster insect populations in order to save the birds. In North America, meanwhile, there is little concern over the demise of a species that is not only an alien but also widely regarded as a threat to native birds. But this attitude assumes that native species would be able to thrive under the nonnative conditions of the modern urban ecosystem. That raises an important question: if house sparrows can't live in our cities, what can?

WATER HYACINTH

Eichhornia crassipes

FOR MORE THAN A CENTURY human beings have been trying to find ways to kill *Eichhornia crassipes*, the lush green free-floating water plant better known as water hyacinth. We've tried poisoning it with herbicides, hacking it to pieces with machetes and overwhelming it with ravenous leaf-eating insects. We've unleashed herbivorous carp fingerlings by the million. We've tried setting it aflame and blasting it to smithereens with explosives. We've used every sort of floating mulching machine that can be imagined — pusher boats, modified backhoes, conveyor harvesters, rakes, shredders — all, as one weed management guide puts it, "specially designed to shear, shred, crush, press, pull, convey, lacerate, and remove aquatic weeds from waterbodies." And the battle continues. One of the most productive plants in nature — and one of the world's worst weeds — water hyacinth remains a pest in the lakes, slow-moving rivers and swamps of more than 50 countries.

It's enough to make a person wonder: isn't there a better way? There may be. Hidden away in the mountain of studies on different ways to destroy water hyacinth is an emerging body of information focused on tackling the problem from a different angle. Rather than viewing water hyacinth as a world threat, this approach looks for ways to use the fast-growing plant to our advantage. What if water hyacinth could be a source of biofuel, thereby preserving valuable croplands and food sources? Perhaps we could feed it to cattle, pigs, ducks and fish. Or maybe this rapidly renewable raw material could become a substitute for wood. Some argue that such talk is dangerous

nonsense that would only make a terrible situation worse. Others, however, see a golden opportunity. If they are right, we may one day be seeing water hyacinth — and perhaps invasive species in general — in a completely different way.

This is not the first time humans have been tempted by water hyacinth. Indeed, one reason we got into this mess is that humans were unable to resist this swamp-dweller's attractions, particularly its exotic beauty and its ability to thrive in different habitats. A native of the floodplains of large South American rivers, water hyacinth sports large, glossy green leaves and attractive light purple flowers, which bloom at the ends of stalks sometimes rising 3 feet (1 m) above the water. After it was brought back to Europe by explorers in the early 1800s (possibly even earlier, according to recent research), water hyacinth became a popular ornamental plant for backyard ponds, botanical gardens, aquariums and even natural lakes. Between the late 19th century and the Second World War it was gradually introduced more widely throughout the tropics and subtropics. In the 1950s and 1960s it was widely distributed in China, as a food source for animals.

Unfortunately, it didn't stop there. Aided by floods carrying it into nearby water systems, additional deliberate introductions, and birds and other vectors transporting its seeds, water hyacinth has continued to spread. Sometimes the infestations are only minor proliferations, but in many places they are followed by periods of explosive growth that leave locals wondering what just hit them. Water hyacinth mats aren't just a few lily pads decoratively dotted here and there. They are floating rafts of vegetation, so densely packed that they can become almost land-like, accumulating wind-blown soil and supporting the growth of other plants. A mat can contain 800,000 plants per acre (two million per ha) and weigh up to 440 tons (400 tonnes), and under optimal conditions it can double in size in six days. At its worst, water hyacinth blankets water surfaces from the shoreline to as far as the eye can see — vast fields of green that can easily be seen from outer space.

↩ This fast-growing, free-floating plant can rise to a height of 3 feet (1 m) above water and proliferate into dense mats containing up to 800,000 plants per square acre (two million per ha).

The impacts of these invasions can be enormous. In parts of the developing world where communities lack resources to mount eradication programs, out-of-control water hyacinth mats have had the effect of a big-city transit strike. Lakes and rivers, frequently vital sources of food, income and transportation, become impenetrable. On the Sepik River in Papua New Guinea, water hyacinth overgrowth essentially barricaded the sole route linking riverside villages with downstream markets and health-care services; several human deaths have been blamed on the problem. The plant's nicknames — beautiful devil, weed from hell — speak volumes. Out-of-control water hyacinth has been blamed for a rise in malaria and other diseases; the mats provide habitat for parasites and viral vectors such as snails and mosquitoes. Hydropower plant maintenance workers have to continually remove the plants in order to prevent them from interfering with operations. And because the plants also lose water through surface evaporation, they're seen as a threat to supplies of fresh water.

Water hyacinth isn't the only invasive plant running wild in freshwater habitats around the world. Since the 1940s the aquatic fern sometimes known as giant salvinia has spread from its native home in South America across Africa, India, the United States, Australia and elsewhere. Other troublemakers on the growing list include hydrilla, Eurasian watermilfoil and water lettuce. However, none — so far, at least — has come close to rivaling water hyacinth, which is frequently referred to as the world's worst aquatic weed.

One secret to the species' success is its hardiness. Although water hyacinth cannot tolerate salt water, currents or even a moderate degree of cold — which generally keeps it out of brackish estuaries and saltwater marshes and makes its year-round presence unlikely north and south of the 40th parallels — it can grow readily under a wide range of warm temperatures and pH and nutrient levels. This, plus its ability to spread by producing seeds and through vegetative reproduction, has enabled it to thrive throughout much of the tropical and subtropical zones.

There's also another important reason for its success. Around the world, most freshwater lakes and rivers have suffered enormous impacts during the past century from increasing human activity. Nutrient-rich runoff from surrounding agricultural lands (exacerbated by deforestation), sewage from settlements, manure from cattle farms and pollution from industries have dramatically changed water chemistry and nutrient levels. These changes have transformed the algae, bacteria and other microbes that make up the base of food webs. Overfishing and introduced exotic fish species have also restructured food webs from the top down. At the same time, dams, irrigation channels, artificial lakes, levees and water diversion systems have completely altered the hydrology (flow rates, flood patterns, water levels) of entire watersheds. In short, freshwater lakes and rivers are now completely different environments than they were in the past. They're now characterized by a unique set of conditions — lack of flooding, high nutrient levels, absence of wetlands, native ecosystems in a state of disarray — that represent the predictable outcome of human presence, regardless of location on Earth. Water hyacinth is perfectly at home in these new environments, as if nature had specially designed it for life under such conditions.

Villagers walk around a boat that has become trapped in a mass of water hyacinth along the Tha Chin River in the Nakhon Pathom province of Thailand. An outbreak of water hyacinth can be particularly hazardous for any community that relies on waterways for food, income or transportation.

This may sound counterintuitive, given what we know about evolution and the way in which species adapt to their native environments. How could a plant from steamy jungle flood zones be super-fit for survival in a world dominated by *Homo sapiens*? Scientists have a good explanation for this paradox, however, one that stems from the highly unusual nature of those floodplains. Home to the world's largest freshwater wetlands, the big river systems of South America — including the Amazon, the Orinoco and the Paraná — wind their way over vast expanses of mostly flat lands. On the plains surrounding the rivers and their tributaries are innumerable temporary lakes that form highly dynamic habitats. During part of the year they are typical lakes, but after periods of heavy rain or when the snow on distant mountains begins to melt, the waters of the nearby rivers rise, swallowing them up. These cyclical fluctuations create a chaotic environment for the organisms in the floodplain lakes. Besides the dynamic hydrological conditions — including water levels that can change by up to 33 feet (10 m) and remain high for months — there are also wide swings in other conditions such as nutrient levels, which rise significantly every time the sediment-rich rivers flood their banks.

Several adaptations appear to have helped water hyacinth become one of the most widespread species across its chaotic native range. As a free-floating plant, it is unperturbed by changing water levels. Perhaps more important, it can capitalize on excessive nutrient loads with explosive growth. This trait may be the key to its success. As newly enriched lakes re-emerge after the floods, water hyacinth is able to outgrow not only other emergent aquatic plants but also submerged vegetation, which ends up being shaded by its floating canopy. However, its life isn't completely ideal. A survey of floodplain lakes associated with the Paraná River found that in high-water periods, water hyacinth dominates, but as the lakes dry out, it slowly succumbs to other species better suited to the conditions of that part of the cycle. The rains invariably return, though, and water hyacinth quickly regains the upper hand. In the Amazon floodplains, similar fluctuations have been observed over longer periods, with water hyacinth flourishing during years of heavy rainfall, when the water is deep or fluctuating, but then giving way to bottom-rooted species during years of lower rainfall.

Given this biology, it's no surprise that water hyacinth has become such a dominant plant. Indeed, water hyacinth has found itself in a new world almost tailor-made to match its competitive strengths. In the process of managing our water systems — damming rivers, creating irrigation systems, containing and diverting flood waters — we've turned most of our freshwater habitats into slow-moving water bodies that are ideal for floating plants. By increasing the amount of nutrients entering these waterways, we've provided resources that water hyacinth can readily exploit to fuel its explosive growth. We've essentially recreated the part of the natural flood cycle in which water hyacinth thrives, only on a permanent basis. And, as in the South American floodplains, no other plant can come close to competing.

All this would explain why water hyacinth is most often a problem in irrigation channels, artificial lakes, reservoirs downstream of dams, rice paddies, and lakes suffering from high loads of nutrients because of deforestation, agriculture and close proximity to human settlements. Indeed, the extent of water hyacinth proliferation often correlates directly with the degree to which a water body has been abused or altered. In Egypt, for example, the plant has been around since the 1880s. It didn't become a big problem, however, until after 1970, when completion of the Aswan High Dam changed the seasonal flow patterns of the Nile River. A similar scenario may also explain why water hyacinth spread widely in California's Sacramento–San Joaquin delta after being present at tolerable levels for more than 30 years.

In many other places the extent of water hyacinth proliferation has been linked directly to the amount of nutrients being channeled into local water systems. A study from India used satellite images taken between 1988 and 2001 to track changes in water hyacinth growth on water bodies near Bangalore. The two lakes that had seen the greatest increase were those that had experienced increasing loads of raw sewage. In general, lakes with high levels of nitrogen, phosphorus and heavy metals plus low visibility were most heavily infested with water hyacinth. Where the water was unpolluted, the invader had failed to grow. "Water quality and climate data," the researchers concluded, "show that high pollution supported robust growth of water hyacinth."

Eradication efforts rarely address this root cause of water hyacinth proliferation, possibly because it is enormously difficult to control the makeup of water entering freshwater systems. Fighting water hyacinth is an often uphill, sometimes never-ending battle, like fighting a fire in the middle of a windstorm. A good example is Lake Victoria in Africa. Water hyacinth didn't make it into the lake's nutrient-enriched waters until 1989, but once it did, the World Bank contributed $8.3 million to an eradication program that included releasing some 142,000 weed-eating weevils. By 2001 the project was being hailed as a remarkable success story. The weevils, perhaps assisted by cooler weather, had reduced the amount of water hyacinth coverage from a high point of nearly 43,000 acres (17,300 ha) early in 1998 to less than 2,500 acres (1,000 ha) by mid-2000. Unfortunately, that didn't solve the problem. Some plants survived, and perhaps more significantly, a large number of seeds — produced in direct response to the stress caused by the weevils — were released into the lake. In 2006, likely encouraged by added nutrients and sediment generated by unusually heavy rains, the plant returned with a vengeance. Although scientists involved in the effort remain confident that the resurgence is part of the cyclical nature of predator-prey relationships, the results illustrate the problems associated with the war against water hyacinth, not just in Lake Victoria but also around the world.

Decades of scientific research on ways to control water hyacinth, coupled with extensive eradication measures involving large amounts of money, time and public education, have succeeded in bringing the problem under control in many places. Unfortunately other areas have not been so lucky; either they lacked the resources to mount effective programs or removal techniques failed because of local circumstances. Even where control measures have succeeded, the threat of reinvasion is always present. The weapons used in this seemingly never-ending battle are obviously less than perfect. Indeed, if the problem has been created by excessive dumping of chemicals and nutrients into the environment, then attacking it with more chemicals and leaving the plants to rot in the water can be seen as a deeply flawed approach.

This brings us to the other approach. Initial attempts to take advantage of this super-weed's remarkable growth and extreme hardiness were focused on the plant's nutritional qualities. Because of the efficiency with which its roots suck nutrients from the water, water hyacinth is a plant loaded with vitamins, minerals, amino acids and protein — all keys to a healthy human diet. Although you'll likely never see water hyacinth in the vegetable section of your local market — for one thing, it doesn't taste very good — it might be able to contribute indirectly to human sustenance. For example, studies from as far back as the 1940s have found that, as a fertilizer, composted water hyacinth carries four times the nutrient punch found in traditional farmyard manure. Alternatively the plant can be processed and fed to animals, as it was in China in the 1950s and 1960s, or used as a source of vitamins and minerals for human consumption. The daily amounts of protein, minerals and vitamins recommended for a healthy adult can be found in less than 7 pounds (3 kg) of water hyacinth leaves. That's a lot of vegetable matter, but in developing countries — where the weed is most prolific — it could be a valuable asset in the war on malnutrition.

What makes the idea even more appealing is that the water hyacinth would also be performing a valuable service to the environment. In the mid-1970s NASA began testing the potential of water plants as natural recycling and purification systems. The original idea was to develop miniature ecosystems for long space voyages, a concept that involved using human waste to grow plants, which would in turn generate food and oxygen to sustain the humans, who would create more waste. The same concept can be applied to communities here on Earth. In this context, plants could act as low-cost filters to rapidly remove nutrients and pollutants from waterways contaminated by sewage and agricultural runoff. The raw material created as the plants became saturated would then be used as fertilizer or as the principal component in methane gas generation.

The NASA program tested various aquatic plants to identify which would be suitable for such uses, and the species that emerged head and shoulders above all the others was water hyacinth. This plant's ability to tolerate high levels of raw sewage, heavy metals and other contaminants meant it could filter even the most polluted water both cheaply and efficiently. Its remarkable growth rate meant it could easily outcompete most other plants in generating biomass. Biochemist Bill Wolverton, who headed the project, calculated that water hyacinth could produce 69 tons of dry plant matter per acre (154 tonnes per ha) each year, more than three times the yield of other candidates such as eucalyptus, sugarcane, sweet sorghum and kelp.

Wolverton developed a water-treatment system consisting of a series of hyacinth-filled pools. Heavily polluted water fed into one end of the system emerged from the other purer than normal tap water. A specially designed harvester removed the saturated plants once a month, after which the rapidly regenerating plants created brand-new filters. As Wolverton imagined it, the harvested plants could be used as an "inexhaustible natural resource" — for fertilizer, poultry feed, an additive to cattle feed, plant mulch, biofuel. There were no limits to the potential of this wonder-weed in his mind. In 1976 he was reported to have said, "I fully intend to solve a major pollution problem, a major energy problem, a major food problem, and a major fertilizer problem."

Obviously Wolverton's ideas have yet to solve these problems, but the idea of a water hyacinth economy has refused to die. Wastewater treatment systems using water hyacinth have been established in Texas (a direct result of Wolverton's efforts) as well as in a number of countries that include France, India, Egypt, China, Argentina and Brazil. At the same time, a growing number of studies have reaffirmed water hyacinth's potential as an efficient waste treatment system, not just for traditional sources such as agricultural runoff and municipal sewage, but also in artificial wetlands created to deal with industrial effluent and in manure ponds created to deal with waste generated by large-scale pig and cattle farms. In particular, water hyacinth's ability to deal with the mess made by giant pig farms in the southern United States is being seen as a potential solution to the increasingly serious problem of groundwater contamination.

Other studies have shown that water hyacinth can successfully treat heavily contaminated effluent generated by some of our dirtiest industries, including textile plants, pulp and paper mills and tanneries. It can eliminate the nitrogen and phosphorus that lead to eutrophication; increase water clarity by removing sediments and microorganisms, including harmful fecal coliform bacteria; and rid water of toxic heavy metals and organic

Don Schumann uses water hyacinth to filter-clean water at his shrimp farm in Vero Beach, Florida. In a growing trend, hyacinth is being used for waste treatment by industries ranging from agriculture to textile factories to pulp and paper mills.

chemicals such as pentachlorophenol (PCP). Researchers have even demonstrated that water hyacinth might be used to cheaply and efficiently recover silver from industrial wastewater. Part of the secret of this natural filtering ability is now known to be electrical charges in water hyacinth's root hairs. These can attract suspended particles, which are then gradually digested and absorbed into the plant.

At the same time, research teams are developing water hyacinth-based biofuel generating systems. In a report published in 2007, researchers from Texas Tech University argued that biomass sources such as water hyacinth are, unlike other alternative energy sources such as solar power and wind, available over a broader geographical area and at any time of day or night. Among the different types of vegetation that could be biomass sources, aquatic plants stand out because they don't require valuable freshwater resources to grow. And then there's the energy punch. According to the Texas Tech study, land plants could produce up to 20 tons per acre (45 tonnes per ha), which would yield 379 million British thermal units (BTUs)

Nutrient-rich, human-modified waters — like this aquatic farm in China's Fujian province — are the prime environment for water hyacinth invasions.

of energy per year. Water hyacinth, with a maximum yield four times greater, could potentially yield one billion BTUs per acre. The researchers calculated that if the United States could use the waste generated by its cattle, pig and poultry farms to grow water hyacinth, the biomass generated each year would amount to more than 80 percent of the current electrical energy needs of the entire country.

Biogas production would require investment in technology that could be prohibitive for developing countries. And many other challenges, such as the high water content of water hyacinth and the cost of transporting it to processing facilities, are no doubt obstacles. But times are changing. The current energy crisis has led the U.S. Department of Energy to set an annual production goal of 60 billion gallons of ethanol by 2030, roughly 30 percent of the country's current gasoline usage. A desire to avoid converting land being used for food production into land devoted to meeting energy needs may drive the development of more efficient technologies for using biomass sources such as water hyacinth.

As for other uses, recent research has demonstrated that surface mulch and compost made from water hyacinth can substantially increase yields of tomatoes and other crops. Throughout China and Southeast Asia, water hyacinth is used as a source of food for fish, ducks, pigs and cattle. In a number of countries small-scale businesses have begun exploiting water hyacinth fibers, which can be turned into rope, baskets, paper and fiberboard for building. You can now order water hyacinth furniture — tables, chairs, even sofas — on the Internet. In India the plant is being used as medicine for treating goiter and to salve wounds. Elsewhere it is being converted into briquettes for cooking, which helps to reduce the deforestation that results when poor villagers are forced to cut down trees for cooking and other fuel needs. Yet more research suggests that water hyacinth has properties that would make it a good source of antioxidants, essential compounds believed to help the body fight cancer and other diseases. Finally, the plant is being tested for use in various industrial processes, including as a source of vitamins in food processing.

To incorporate productive uses into the control of water hyacinth on a large scale will require social and cultural shifts that aren't easily made. Apart from the initial technological hurdles, you need coordinated harvesting, hauling, processing, production and market development before any part of the system can begin to function. And for this to occur you need commitment from policymakers, community organizations and private investors. This first step may be the hardest, given water hyacinth's reputation. As one scientist recently remarked, "It is very difficult to answer the question, Is *E. crassipes* the golden plant, or the world's worst aquatic weed?"

One last point in the debate is what would happen if we did win the war on water hyacinth. While successful eradication efforts have reduced the hardships caused by severe outbreaks (as on the Sepik River), it can also result in unintended consequences. One example is the Garças Reservoir in São Paulo, Brazil. Since the mid-1950s the reservoir had been increasingly a dumping site for sewage and other sources of nutrients, which eventually triggered an outbreak of water hyacinth. In 1997 the plant blanketed between 10 and 20 percent of the water surface. A year later, when the spread was at its peak, nearly three-quarters of the reservoir was covered. Local authorities, responding to an outbreak of mosquitoes associated with the invasion, began a program of mechanical removal that took two months to complete. While the effort helped reduce mosquito levels, it also sent the reservoir's ecosystem into a downward spiral. Without water hyacinth to sop up the nutrients, increases in phosphorus and nitrogen levels resulted in an explosive outbreak of cyanobacteria — including potentially toxic *Microcystis aeruginosa* — that persisted throughout a five-year study period.

The researchers suspected that the reservoir's response was a case of what ecologists call "abrupt permanent impact." Their theory is that the system degenerated into a simple cycle in which rapidly reproducing cyanobacteria filled the reservoir, died and then settled to the bottom, triggering the release of more nutrients and growth of more cyanobacteria. The persistent mass of microbes also blocked sunlight and reduced oxygen levels, two factors associated with aquatic "dead zones." As the researchers concluded, "Particularly after the water hyacinth harvest, internal feedback mechanisms took over, amplifying the auto-fertilisation process, and leading to an irreversible jump to a more turbid and degraded state."

A similar outcome was recorded in reservoirs in Mexico after water hyacinth removal, but an even more interesting case comes from Kings Bay, Florida, a spring-fed inland bay that forms the headwaters of the Crystal River and is an important wintertime habitat for the endangered West Indian manatee. A team of scientists from the University of Florida found that the bay's relationship with water hyacinth followed the familiar routine. Up until the 1950s it had been present but not a nuisance. Then the area went through changes that included loss of wetlands, widespread construction of ditches and canals, commercial and residential development of land throughout the watershed, and deterioration of water quality. Water hyacinth plants began to proliferate, until an aerial herbicide bombardment not only killed the invader but also turned once stunningly clear waters into a cesspool of mud and dying plants.

From the early 1960s the bay was then overrun by hydrilla, an even more troublesome aquatic invader. Despite hindering navigation, it is thought to have improved water clarity and bolstered the manatee population, providing fodder for these large sea mammals. However, it was an unwelcome invasive species. First hundreds of thousands of gallons of sulfuric acid were dumped in the bay, and later a combination of herbicides. The hydrilla simply died off and then bounced back, while the manatees' organs began to accumulate heavy metals. In the fall of 1985 nature intervened in the form of Hurricane Elena; it is believed to have flushed in seawater that killed off the hydrilla far more effectively than any poison.

Following the elimination of hydrilla, the bay's waters were taken over by large-scale cyanobacteria outbreaks. The waters now resemble a slimy, smelly soup. Worse, the dominant bacteria have locked up the bay's nutrients in a form that cannot be eaten by the manatees, and they may even be toxic. Battle lines have been drawn between aquatic-plant managers, whose job it is to control invasive species and restore native ecosystems, and the local residents, many of whom benefit from tourism generated by the manatee population. After attempts to filter the cyanobacteria failed, in 2006 the plant-control authorities proposed all-out chemical warfare: a potent combination of plant- and microbe-killing agents. But some local residents think the best answer may be to work with nature rather than against it. To this end, they want to bring back water hyacinth.

Driven by recollections of the bay during the days when water hyacinth was present, they've recently suggested allowing the plant to grow in a controlled manner. The idea is that water hyacinth will out-duel the cyanobacteria for nutrients, as it has done elsewhere, purifying the water while at the same time providing food for manatees and possibly even helping restore the health of native submerged plants. When University of Florida researchers asked various wildlife managers about the Kings Bay situation, they found some who agreed with the proposal. But most of them were appalled, and the managers who currently control the fate of Kings Bay were equally dismissive of the idea. As the case studies illustrate, certain attitudes toward nonnative species are deeply entrenched. Water hyacinth may one day change the way we think about invasive plants and animals. That day, however, has yet to come.

NILE PERCH

Lates niloticus

Y**OU CAN READ ABOUT** Nile perch in almost any textbook dealing with conservation biology, ecology or invasive species. Evolutionary biologists are all too familiar with the tale. It's a story about an amazing assemblage of colorful African fish known as cichlids and the large, ravenous predator that was introduced into their midst. The big fish ate all the little fish, and the little fish became extinct. The end.

That's the short version, anyway. The reality is a great deal more complex. Indeed, some 15 years after the book was supposed to have been closed on the Nile perch, the saga continues to generate new twists and turns that remain less widely chronicled. What we now have is an increasingly ambiguous narrative in which the villain is actually a hero (at least to some) and the victims aren't in fact dead (at least not all of them).

First, let's meet the antagonist: *Lates niloticus*, the Nile perch. One of the world's largest freshwater fish species, *L. niloticus*'s native range includes the Nile River and many of its tributaries, as well as several other major river systems and lakes throughout Africa between the Sahara and the equator. Its body is silvery gray with tinges of blue, and its distinctive eyes resemble jet black pearls surrounded by bright yellow discs. While many predatory fish are sleek and streamlined, a Nile perch has a certain ugliness; its lumpy body brings to mind a slightly herniated blow-up beach

◀ Nile perch can grow to lengths exceeding six feet (2 m) and weights of close to 440 pounds (200 kg). It's no surprise that in the decades following its introduction some villagers in Uganda feared the species was capable of eating children.

toy. Its most striking feature, however, is its size. Adults can grow to be more than 6 feet (2 m) long and weigh up to 440 pounds (200 kg). String up a giant Nile perch beside a full-grown man and it isn't obvious which would emerge the victor in a fair fight.

Our protagonists are the Lake Victoria cichlids. Although not found in North America or Europe, cichlids (pronounced *SIK-lids*) are in fact one of the most diverse groups of freshwater fishes. By one recent estimate there are about 3,000 known species, most of them found in the warmer parts of South America and Africa. New members are still being discovered, however, and scientists estimate that the total number may be much higher. These fish come in a wide variety of shapes and sizes, but often their most interesting features are the bright colors and body markings that characterize many members of the group. With their rich reds, vivid yellows and metallic blues, as well as myriad intricate patterns, many cichlids would not look out of place around a coral reef. Aquarium enthusiasts love them.

Finally, there's the setting: Lake Victoria. At close to 27,000 square miles (nearly 70,000 sq km) it's almost the size of Ireland, making it Africa's largest lake and the biggest expanse of fresh water found anywhere in the tropics. It lies in the western part of the Great Rift Valley, its shorelines shared among Uganda, Tanzania and Kenya.

Our story begins some 14,700 years ago, when this portion of East Africa was emerging from a 2,500-year dry spell and low-lying inland basins had filled with water. No one knows whether only one ancestral cichlid species made its way into newly formed Lake Victoria, or several, nor do they know how these original fish got there. It is suspected that the founders, with remarkable speed, underwent an explosive burst of what is known as adaptive radiation. In just a few short millennia — a blink of an eye compared to the rest of life's long evolution — these tiny fish diversified into distinct species that invaded every ecological nook and cranny, from the deeper regions of the lake's interior to the wetlands and rocky shorelines of its edges. In many places numerous species developed, all from the same genus, and each with its own distinctive markings and color patterns and its own non-interbreeding population. By modern times an estimated 500 different species represented 80 percent of the lake's total fish biomass. Many of these species could be found nowhere else.

Fast forward to 1954. East Africa was in the twilight of the colonial period and the British authorities were enacting policies aimed at promoting self-sufficiency among their subjects. It's also possible that more than a little self-interest was involved, as certain administrators did enjoy big game hunting and fishing. Either way, the result was the introduction of Nile perch into several large African lakes, including Lake Victoria, first in a clandestine operation believed to have occurred in 1954, and then in a more organized manner at least twice more in the early 1960s. There was a clear rationale behind these actions. When the British began penetrating deeper into East Africa in pursuit of resources during the late 1800s, the Lake Victoria basin was still remote and sparsely populated. With the construction of railway lines and the introduction of modern fishing gear and techniques, however, the area gained new importance as a fishery. This had predictable consequences: by the 1920s the principal fish stocks, including a highly prized tilapia species known locally as *ngege* (which belongs to the cichlid group), were already in decline.

This wasn't much of a problem for local fisherman, who, according to research by ecologist Robert Pringle, continued to exploit the lake's smaller species. To the British, however, it was a wasted opportunity. Yes, the lake was teeming with fish, but they were little fish, "trash fish" that required a lot of time and effort to catch and process — too many bloody bones and not enough flesh! At one point it was suggested that cichlids should be scooped from the lake and dumped onto croplands as a cheap form of fertilizer. But others believed an even better move would be introduction of a major predator that could more easily be exploited by a commercial fishery. By swimming around and eating up all the small fish, such a predator would be converting worthless biomass into big, meaty fillets. Some ecologists feared that such a move would have disastrous consequences, but this was the 1950s. The hard lessons of species introductions — particularly those involving top predators — had yet to be widely appreciated. The warnings were ignored and the Nile perch, along with four exotic species of tilapia, was introduced.

At first it wasn't clear whether the plan was going to be the success that proponents had hoped or the disaster that some scientists feared. Indeed, during the 20 years that followed the introductions, Nile perch were caught only in small numbers in a few locations. But then something happened. In the late 1970s the population of Nile perch exploded, and almost overnight a new industry emerged on the back of this easily exploitable and increasingly lucrative resource. Dozens of fish processing plants were erected around the lake, and people from surrounding regions flocked to the growing lakeside communities to participate in the take. And what a take it was — during the early 1990s, according to George Barlow, author of *The Cichlid Fishes*, Lake Victoria became the most productive freshwater lake fishery in the world, thanks mainly to a Nile perch harvest estimated at between 200,000 and 300,000 tons per year.

The enormous haul was being driven by a growing demand for Nile perch overseas. Before long, Nile perch fillets were being shipped all around the world, as well as to other parts of Africa. Wealthy sports fishermen began flying in from Europe and North America, lured by the notion of landing a 400-pound lake fish. The impacts were dramatic. In Kenya alone the number of fishermen on Lake Victoria rose from 11,000 in 1971 to nearly 40,000 by 2001.

But the affair quickly came to be seen by many observers as one enormous blunder. As local communities began experiencing problems frequently associated with rapid development — the deterioration of traditional culture and social systems; increases in crime, prostitution, urban squalor and alcohol and drug abuse; unequal distribution of wealth and monopolization of the resource by a powerful elite — the Nile perch quickly became a symbol for those who saw such initiatives as yet another form of Western imperialism. There were also some completely unforeseen consequences. Before the Nile perch boom, for example, local grass-smoking techniques were used to cure the smaller native fish. Preservation of the bigger, oilier perch, however, required firewood. This, along with an increase in wood demand for boat building, housing construction and other activities associated with the boom, resulted in rapid decimation of the local forests.

Human-driven impacts have transformed the ecology of Lake Victoria in Africa. Pollution and eutrophication are likely at least partly responsible for the decline of native species in the lake.

Ecologists and evolutionary biologists were also alarmed by the rise of the Nile perch, but for different reasons. When researchers surveyed the lake in the late 1980s and early 1990s, they discovered that while the nonnative species had been booming, populations of many native cichlid species had gone into a steep decline. Some 200 of Lake Victoria's celebrated cichlids were declared extinct — "an episode of mass extinction," according to one report, "rarely paralleled in history."

Harmful invasions always attract attention, but in this case the negative feelings were particularly acute. In addition to making the International Union for Conservation of Nature's Top 100 most dangerous aliens list, the Nile perch was the inspiration for a powerful documentary by European director Hubert Sauper, titled *Darwin's Nightmare*. Released in 2004, the film depicted in graphic detail the tragic ills that had resulted from the Nile perch boom: glue-sniffing homeless children, rampant prostitution and widespread proliferation of AIDS and poverty. Particularly appalling were scenes showing locals collecting the severed fish heads lined up outside processing plants; they could no longer afford to buy the rest of the fish because overseas demand had pushed its price beyond their reach. The film also insinuated that cargo planes exporting Nile perch to Russia were returning laden with arms, the implication being that, in addition to everything else, this invasive species was also fueling ongoing war throughout the region. The film's gripping images, not to mention its Academy Award nomination, left little room for ambiguity, serving as what seemed to be the final word on an unspeakable tragedy.

But during the past decade, mounting evidence suggests that the case is not nearly so clear-cut as was widely thought. There is heated debate over whether the impacts of the Nile fish boom have been as severe as they've been depicted. Indeed, many argue that by creating an industry that generates as much as a quarter of a billion dollars a year in foreign currency and has created thousands of jobs, the arrival of the Nile perch has been a huge blessing for a region not overly endowed with income-generating opportunities. And the downside may have been exaggerated as well. Robert Pringle found that local fishermen had indeed been perturbed by the appearance of Nile perch — they found its smell unpleasant, the big fish tore nets that had been designed for catching smaller, less aggressive species, and some locals believed the fish caused disease, leading to rumors that it was part of a British plot to kill Africans. There were even fears that *mputa* (as the species is known in Uganda) was capable of eating children. However, perhaps the ultimate beef was the lack of demand. In the early years there was no market for the fish, which meant that it was essentially a worthless commodity.

Within a few years, however, most of the reluctance had switched to widespread acceptance. People grew accustomed to the taste, partly, it's been said, because the fish became less oily over time as their prey base diminished. The fishermen invested in new gear and adopted new techniques. But the biggest change was the emergence of a huge, lucrative new market. Indeed, as demand for Nile perch rose, locals began referring to it as "gold" and "savior." And while no one disputed the bleak images of *Darwin's Nightmare*, there were also uplifting stories of local fishermen who could now afford to send their children to school. According to Pringle, "The development of this industry was a tremendous boon for the national and local economies of Kenya, Tanzania and Uganda." To understand just how big, consider this: in 1996 Uganda — a country without a sea coast — earned US$46 million from the sale of 16,400 tons of fish, mostly Nile perch. Among the country's export commodities, only coffee generated more income.

As for the charge that locals are eating less fish because of its high price, driven by overseas demand, a survey published in 2008 found that people weren't shunning Nile perch because they couldn't afford it but because it was worth more to them as a source of cash. As the authors of the study put it, "Fishermen *choose* to sell their fish to fish factories" (their emphasis). "Virtually everyone consulted" agreed that the Nile perch export trade has had a positive impact on lakeside communities.

In terms of ecological consequences, various studies now support the notion that the relationship between Nile perch and the disappearance of the cichlids may not be nearly as straightforward as commonly thought. For one thing, as some researchers have long pointed out, the rapid rise of Nile perch wasn't the only environmental change to have coincided with the decline of the cichlids. Well before the invader became a prominent local presence, decades of deforestation and conversion of land from forest and wetlands to farms and settlements had greatly increased the amount of sediments and nutrients from fertilizer being washed into the lake with each rainstorm. One effect was the onset of eutrophication, in which excessive nutrient loads drive the growth and decay of certain bacteria and plant life, which in turn leads to depletion of oxygen in the water.

Analysis of sediment samples from the lake revealed that during the 20th century there was a sudden restructuring of the lower levels of Lake Victoria's food chain. Once based predominantly on single-celled organisms known as diatoms, it became dominated by nutrient-loving toxic cyanobacteria and eutrophic conditions. According to one recent estimate, up to 50 percent of the lake's potential fish habitat zones have been lost because there's no longer enough oxygen, and it has been suggested that this may have been a contributing factor in the decline of the lake's native species. What's more, the shift occurred in the late 1960s and early 1970s, right before the cichlid populations began disappearing.

At the same time, the increases in algae and sediments and other forms of pollution have combined to reduce water clarity in many parts of lake. In the nearshore areas, for example, the water visibility went from nearly 10 feet (3 m) in the late 1980s to just 5 feet (1.5 m) in the late 1990s. This is thought to have contributed to the cichlids' woes to an even greater degree, because clear water has been identified as key to their evolution and ongoing survival. For example, lakes with better visibility tend to have a greater diversity of cichlid species compared to murkier bodies of water, and within Lake Victoria more species cluster in parts of the lake where visibility is greatest. This is thought to be because cichlids depend heavily on vision for their survival. They use their eyes to hunt food and they rely on vision to select their mates — the bright colors and flamboyant patterns of many male cichlids evolved as a form of visual advertising. This suggests that the establishment and maintenance of distinct species has long depended on females' being able to readily recognize members of their own clan. It also suggests that the recent changes to the water in Lake Victoria may have contributed to the cichlids' decline. Further evidence comes from a study in the late 1990s in which researchers put female cichlids together with males from different species under different lighting conditions. They discovered that when a female could easily see the other fish, she invariably mated with a male from her own species. When the water was murky, her mate selection was completely random.

Yet another factor may have contributed to cichlid decline: overfishing. In 2008 scientists from the Netherlands and Tanzania published the results of a study in which they compared historical fishing reports from various parts of Lake Victoria. They show that catches of both large and small cichlids began to decline not with the appearance of large numbers of Nile perch in the 1980s, but after 1973, when locals first began fishing with bottom trawlers. The authors argue that the predator was still a major factor in the disappearance of the cichlids, citing the continued decline of native fish even in local areas once bottom trawling was discontinued, as well as the disappearance of cichlids from parts of the lake that had not been heavily fished.

The study also points out that a major piece of the puzzle must still be missing. It took 25 years for the Nile perch population to explode, despite the fact that these predators had been swimming around in waters teeming with prey. What triggered the sudden outburst? According to the Dutch scientists, a possible explanation is that during the early years after the Nile perch introductions, the relationship of predator and prey was the opposite of what one would have expected. The large cichlids may have been eating Nile perch eggs and larvae, as well as outcompeting young perch in the battle over plankton and insect larvae. But then, according to this hypothesis, bottom trawling turned the tables. As large cichlids declined in heavily fished areas, more Nile perch would have survived to become big enough to gobble up their smaller rivals. As the cichlid population declined still further, large numbers of increasingly hungry perch would have migrated to other parts of the lake.

While there is little doubt that Nile perch have had a huge impact on the native fish of Lake Victoria, the effects of overfishing, eutrophication and decreased water clarity, combined with clearing of vegetation associated with swamp habitat, make it clear that another invasive species — *Homo sapiens* — has had a just as great or even greater impact on the Lake Victoria ecosystem. Which brings us to our latest chapter. After reaching a peak in 1990, total annual Nile perch catch rates have been in decline and the larger fish have become rare — two classic signs that the population is being overfished. At the same time there has been a resurgence of many native fish, including a handful of cichlid species previously thought to have become extinct. Researchers have discovered, however, that many of these cichlids appear to be hybrids, indicating that poor water quality may be continuing to interfere with cichlid mating patterns.

If there's a moral here, it remains obscure. On the one hand, the introduction of the Nile perch and the likely extinction of up to 40 percent of the lake's cichlid species have not led to a collapse of the Lake Victoria ecosystem (at least not yet). In fact, one could argue that it has expanded to include two new tiers: the predatory fish and a much larger number of human consumers. On the other hand, hopes for this new system's achieving any sort of stability seem dim in light of growing human populations, unsustainable harvests and poor environmental stewardship. Thus, whether the invasion

CONCLUSION

I T'S BECOMING CLEAR that invasive species, long viewed solely as isolated destructive forces, are also creative forces that are capable of positive interactions with other species, including natives. Such interactions are the foundation of ecology — they are what ecosystems are based upon. And given that invasion is happening most intensively in environments that have been overturned by human activity, the emergence of new communities of organisms — including species both native and not — can be viewed as nature's attempt to start over, to rebuild ecological relationships in places where ecosystems have been shattered by impacts such as overhunting, land clearing and pollution.

Some ecologists are now trying to convince others that it's high time we recognized invaded habitats as ecosystems so that research funds can be directed toward studying how these ecosystems function. There is a sense of urgency because even though the evidence suggests that emerging ecosystems are vibrant and diverse, they are also novel. No one knows yet whether the Baltic Sea or the forests of Puerto Rico or any of the other emerging ecosystems on the planet are capable of performing the functions we have come to expect our ecosystems to perform — and on which we depend. This is a key point. If you have the right communities of organisms, then you have systems in place for purifying air and water, cycling nutrients, recycling waste, creating fertile soil, sustaining diverse food chains, preventing the erosion of soil, buffering climate swings, limiting pest and disease outbreaks, facilitating retention of water and converting energy into organic matter. It should go without saying that humans, like every other species with which we share our environment, depend on such ecological processes for survival.

It may be that these new ecosystems will indeed provide us with resources that will benefit our health, happiness and well-being — fresh water, food and nice places to visit. By mitigating the impacts of human environmental abuse, they may even end up playing a crucial role in saving us from ourselves. In the Baltic, for example, most of the ecological changes are taking place in the sea's southern habitats, where environmental abuse by humans has been most severe. Perhaps the blossoming ecosystems there should be viewed as nature's repair crew, "a coping mechanism with which a stressed system maintains its most important ecological functions." One scientist optimistically states: "I expect novel forests to behave ecologically as native forests do, i.e., protect soil, cycle nutrients, support wildlife, store carbon, and maintaining watershed functions." But there's also a chance that emerging ecosystems won't be able to buffer outbreaks of disease-causing microbes, to control the spread of toxic life-forms or to maintain air, water and soil conditions within the bounds of what's considered necessary for human survival.

For those who see no good in the rise of invasive species, there is only one road to take. We must stop alien species from spreading and we must do whatever we can to limit what might be viewed as a rapidly spreading cancer within the biosphere. We can't afford caution. At stake are biological resources that are a legacy of the geological time scale — species and ecosystems that have evolved and formed partnerships over the course of thousands if not millions of years, precious resources that, once lost, will be lost for good. But opponents say this approach is flawed on many levels. For one thing, eradication campaigns are enormous drains on labor and money. Second, they often result in unforeseen consequences and generally fail to address the underlying causes, such as environmental degradation, that turn habitats into invasion magnets in the first place.

Finally and perhaps most damningly, opponents of the war on invasive species contend that it has lost its way because it's being driven by ideas rooted not in science but in culturally constructed notions about what nature should be. If the goal is to return an environment to its native state, what exactly does that mean? Was it the state that existed prior to European contact? And which species should we deem "good" and which ones get labeled "bad"? Should the cattle egret — an African species that made its way across the Atlantic on its own and has spread dramatically throughout the Americas during the past century — now be viewed as a native or an invader? And what about the eastern barred owl, which has taken advantage of human habitat changes to spread into the forests of western North America? Should it be granted amnesty or sentenced to extermination? Should the many Brits who enjoy carp fishing care that their pleasure derives from aliens introduced two millennia ago by the Romans? There are no correct answers, because such questions mean different things to different people depending on their personal ideas of what's good and what's not.

A more scientific approach would be to look beyond the arbitrary oppositions of native versus nonnative and good versus bad and accept that in addition to causing harm, introduced species may also be capable of improving ecosystem health, particularly in environments that have been sickened by human abuse. Currently there is a heavy bias against alien species in the supposedly unbiased reports published by invasion biologists. One can find invasive species being accused of ecological impacts such as fixing nitrogen, circulating nutrients and removing sediments from water, even though ecologists have long pressed for preservation of native species primarily because they provide exactly the same services. (As Mark Sagoff, a frequent critic of the war on invasive species, recently put it, "Were it native, the zebra mussel might be hailed as a savior, not reviled as a scourge.")

These biases are further amplified in the media, as reporters are eager to fuel the public's lust for a good horror story. Most people have heard about the spread of purple loosestrife and the hundreds of millions of dollars invested in halting its takeover of North American wetlands. But few have heard about recent research that suggests it hasn't caused any major ecological damage at all, and that it actually supports an abundance of native bees and other insects. Many people have cringed at alarming headlines about the spread of Japanese honeysuckle. But how many have been informed that this vine serves a number of valuable ecological functions, including nutrient retention, carbon storage and habitat improvement for wildlife?

This is not to say that invasive species don't cause trouble. The worst examples are the disease-causing microbes that with increasing regularity threaten both our factory farms and our densely populated and frequently intermingling societies, and the weeds and insects that threaten our monoculture crops and managed forests. We can blame invasive species for the social changes brought about by the Irish potato famine (triggered by a fungus introduced from North America) and bubonic plague (caused by invasive bacteria lurking inside invasive fleas riding aboard invasive rats), as well as the epidemics that are no doubt to come. And there's no denying that invasive species — particularly in their early stages of proliferation — can alter habitats in ways that cause major headaches for people trying to use boats, engage in leisure pursuits or operate businesses.

A more toned-down view of invasions does not give humans license to continue with business as usual. To suggest that nature can respond in certain positive ways to past environmental abuse is not to say that nature will be able to do so indefinitely. Destruction and alteration of habitat continue to be the major threat to biodiversity. Nowhere in the more moderate view of invasive species is there any indication that continued human population growth and loss of native habitat won't result in another mass extinction event. After all, the forests of Puerto Rico may be filled with invasive species, but they wouldn't be forests at all if the land had remained under cultivation. Similarly, if we continue to harvest large marine predators at unsustainable rates, then tuna and swordfish and their like will go the way of the bison and sturgeon. During the past century the amount of carbon dioxide in the atmosphere has increased by a hundred parts per million, and is now reaching levels that are thought not to have existed in the past 400,000 years. The drop in the pH of seawater, which began declining at the start of the Industrial Revolution, has occurred a hundred times faster than any similar shifts in ocean chemistry over the past 20 million years. If we want to go on toying with the amount of carbon dioxide in the atmosphere or the pH of the oceans or the mercury levels in our soil . . . well, good luck to us all.

Looking at invasive species in a more positive light opens our eyes to the fact that nature exists beyond the boundaries of the untouched rain forest, the untrampled alpine meadow and the pristine lake. This should come as a welcome revelation, because although there are undeniable merits in preserving what remains of Earth's rapidly shrinking undisturbed environments, the reality is that the human footprint — which includes the reshuffling of life-forms, the alteration of habitats and the rise of new ecosystems dominated by nonnative species — is now enormous and intractable. Making sure these environments remain livable may turn out to be one of the biggest challenges we've ever faced. But if we accept that the forces of ecology are constructive as well as destructive, we just may find ourselves surrounded by a surprisingly long list of allies.

REFERENCES

Part One: NATURE RESHUFFLED

Minchin, D., S. Gollasch and I. Wallentinus. *Vector Pathways and the Spread of Exotic Species in the Sea*. ICES Cooperative Research Report No. 271. Copenhagen: International Council for the Exploration of the Sea, 2005.

Pimentel, D., R. Zuniga and D. Morrison. "Update on the Environmental and Economic Costs Associated with Alien-Invasive Species in the United States." *Ecological Economics* 52 (2005): 273–88.

1   AMERICAN BULLFROG *Lithobates catesbeianus*

D'Amore, A., V. Hemingway and K. Wasson. "Do a Threatened Native Amphibian and Its Invasive Cogener Differ in Response to Human Alteration of the Landscape?" *Biological Invasions* 12 (2010): 145–54.

Ficetola, G. F., et al. "Pattern of Distribution of the American Bullfrog *Rana catesbeiana* in Europe." *Biological Invasions* 9 (2007): 767–72.

Garner, T. W. J., et al. "The Emerging Amphibian Pathogen *Batrachochytrium dendrobatidis* Globally Infects Introduced Populations of the North American Bullfrog, *Rana catesbeiana*." *Biology Letters* 2 (2006): 455–59.

2   BROWN TREE SNAKE *Boiga irregularis*

Rodda, G. H., T. H. Fritts and D. Chiszar. "The Disappearance of Guam's Wildlife." *Bioscience* 47 (1997): 565–74.

3   CHYTRID FUNGUS *Batrachochytrium dendrobatidis*

Berger, L., et al. "Chytridiomycosis Causes Amphibian Mortality Associated with Population Declines in the Rain Forests of Australia and South America." *Proceedings of the National Academy of Sciences* 95 (1998): 9031–36.

Blaustein, A. R., and J. M. Kiesecker. "Complexity in Conservation: Lessons from the Global Decline of Amphibian Populations." *Ecology Letters* 5 (2002): 597–608.

Di Rosa, I., et al. "The Proximate Cause of Frog Declines?" *Nature* 447 (2007): E4–5.

Krajick, K. "The Lost World of the Kihansi Toad." *Science* 311 (2006): 1230–32.

Skerratt, L. F., et al. "Spread of Chytridiomycosis Has Caused the Rapid Global Decline and Extinction of Frogs." *EcoHealth* 4 (2007): 125–34.

Speare, R., and L. Berger. "Global Distribution of Chytridiomycosis in Amphibians," November 11, 2000. http://www.jcu.edu.au/school/phtm/PHTM/frogs/chyglob.htm (updated April 14, 2004).

4   HUMBOLDT SQUID *Dosidicus gigas*

Masters, Ryan. "The Vicious Giant Squid *Dosidicus gigas*, the Humboldt Squid, Seems to Be Finding a New Home Right off Our Shores." *Monterey County Weekly*, March 10, 2005. http://montereycountyweekly.com.

Zeiberg, L. D., and B. H. Robinson. "Invasive Range Expansion by the Humboldt Squid, *Dosidicus gigas*, in the Eastern North Pacific." *Proceedings of the National Academy of Sciences* 104 (2007): 12,948–50.

5   EUROPEAN GREEN CRAB *Carcinus maenas*

Freeman, A. S., and J. E. Byers. "Divergent Induced responses to an Invasive Predator in Marine Mussel Populations." *Science* 313 (2006): 831–33.

Part Two: EQUILIBRIUM LOST

King, J. R., and W. R. Tschinkel. "Experimental Evidence that Human Impacts Drive Fire Ant Invasions and Ecological Change." *Proceedings of the National Academy of Sciences* 105 (2008): 20,339–43.

6   EASTERN GRAY SQUIRREL *Sciurus carolinensis*

Newson, S. E., et al. "Potential Impact of Grey Squirrels *Sciurus carolinensis* on Woodland Bird Populations in England." *Journal of Ornithology* 151 (2010): 211–18.

Spencer, Ruston. "Who Let the Squirrels Out?" Rushton Spencer blog, September 27, 2008. http://rushtonspencer.wordpress.com.

Tibbetts, Graham. "Sterilisation Plan for Grey Squirrels." Telegraph.co.uk, October 2, 2007. http://www.telegraph.co.uk.

7   KILLER ALGAE *Caulerpa taxifolia*

Occhipinti-Ambrogi, A., and D. Savini. "Biological Invasions as a Component of Global Change in Stressed Marine Ecosystems." *Marine Pollution Bulletin* 46 (2003): 542–51.

8   FERAL PIG *Sus scrofa*

Engeman, R. M., et al. "An Extraordinary Patch of Feral Hog Damage in Florida Before and After Initiating Hog Removal." *Human–Wildlife Conflicts* 1 (2007): 271–75.

Ickes, K., C. J. Paciorek and S. C. Thomas. "Impacts of Nest Construction by Native Pigs (*Sus scrofa*) on Lowland Malaysian Rain Forest Saplings." *Ecology* 86 (2005): 1540–47.

Kotanen, P. M. "Responses of Vegetation to a Change Regime of Disturbance: Effects of Feral Pigs in a California Coastal Prairie." *Ecogeography* 18 (1995): 190–99.

Kristiansson, H. "Crop Damage by Wild Boars in Central Sweden." In *Proceedings of the XVIIth Congress of the International Union of Game Biologists,* 605–9. Brussels: International Union of Game Biologists, 1985.

Kuiters, A. T., and P. A. Slim. "Regeneration of Mixed Deciduous Forest in a Dutch Forest-Heathland, Following a Reduction of Ungulate Densities." *Biological Conservation* 105 (2002): 65–74.

Schley, L. S., et al. "Patterns of Crop Damage by Wild Boar (*Sus scrofa*) in Luxembourg over a 10-Year Period. *European Journal of Wildlife Research* 54 (2008): 589–99.

9   GIANT AFRICAN LAND SNAIL *Achatina fulica*

Civeyrel, L., and D. Simberloff. "A Tale of Two Snails: Is the Cure Worse than the Disease?" *Biodiversity and Conservation* 5 (1996): 1231 52.

Mead, A. R. "Economic Malacology with Particular Reference to *Achatina fulica*." In *Pulmonates*, vol. 213, edited by V. Fretter and J. Peake. London: Academic Press, 1979.

———. *The Giant African Snail: A Problem in Economic Malacology*. Chicago: University of Chicago Press, 1961.

10   C. DIFFICILE *Clostridium difficile*

Dumford, D. M., et al. "What Is on That Keyboard? Detecting Hidden Environmental Reservoirs of *Clostridium difficile* During an Outbreak Associated with North American Pulsed-Field Gel Electrophoresis Type 1 Strains." *American Journal of Infection Control* 37 (2009): 15–19.

Freter, R. "The Fatal Enteric Cholera Infection in the Guinea Pig, Achieved by Inhibition of Normal Enteric Flora." *Journal of Infectious Diseases* 97 (1955): 57–65.

Jernberg, C., et al. "Long-Term Ecological Impacts of Antibiotic Administration on the Human Intestinal Microbiota." *ISME Journal* 1 (2007): 56–66.

Tannock, G. W. *Normal Microflora: An Introduction to Microbes Inhabiting the Human Body*. London: Chapman and Hall, 1995.

Viscidi, R., S. Willey and J. G. Bartlett. "Isolation Rates and Toxigenic Potential of *Clostridium difficile* Isolates from Various Populations." *Gastroenterology* 81 (1981): 5–9.

Walk, S. T., and V. B. Young. "Emerging Insights into Antibiotic-Associated Diarrhea and *Clostridium difficile* Infection through the Lens of Microbial Ecology." *Interdisciplinary Perspectives on Infectious Diseases* (2008). http://www.hindawi.com/journals/ipid/2008/125081.html.

Wilson, K. H. "The Microecology of *Clostridium difficile*." *Clinical Infectious Diseases* 16 (1993): S214–18.

Young, V. B., and T. M. Schmidt. "Antibiotic-Associated Diarrhea Accompanied by Large-Scale Alterations in the Composition of the Fecal Microbiota." *Journal of Clinical Microbiology* 42 (2004): 1203–6.

Part Three: ARRIVAL OF THE FITTEST

Lavergne, S., and J. Molofsky. "Increased Genetic Variation and Evolutionary Potential Drive the Success of an Invasive Grass." *Proceedings of the National Academy of Sciences* 104 (2007): 3883–88.

Maron, J. L., et al. "Rapid Evolution of an Invasive Plant." *Ecological Monographs* 74 (2004): 261–80.

Petran, K., and T. J. Case. "An Experimental Demonstration of Exploitation Competition in an Ongoing Invasion." *Ecology* 77 (1996): 118–32.

———. "Habitat Structure Determines Competition Intensity and Invasion Success in Gecko Lizards." *Proceedings of the National Academy of Sciences* 95 (1998): 11,739–44.

11   ARGENTINE ANT *Linepithema humile*

Holway, D. A., A. V. Suarez and T. J. Case. "Role of Abiotic Factors in Governing Susceptibility to Invasion: A Test with Argentine Ants." *Ecology* 83 (2002): 1610–19.

12   CROWN-OF-THORNS STARFISH *Acanthaster planci*

Brodie, J. E. "Enhancement of Larval and Juvenile Survival and Recruitment in *Acanthaster planci* from the Effects of Terrestrial Runoff: A Review." *Australian Journal of Marine and Freshwater Research* 43 (1992): 539–54.

Brodie, J., et al. "Are Increased Nutrient Inputs Responsible for More Outbreaks of Crown-of-Thorns Starfish? An Appraisal of the Evidence." *Marine Pollution Bulletin* 51 (2005): 266–78.

DeVantier, L. M., and T. J. Done. "Inferring Past Outbreaks of the Crown-of-Thorns Seastar from Scar Patterns on Coral Heads." In *Geological Approaches to Coral Reef Ecology*, part 2, edited by Richard B. Aronson, 85–125. New York: Springer, 2007.

Sweatman, H. "No-Take Reserves Protect Coral Reefs from Predatory Starfish." *Current Biology* 18 (2008): R598–99.

Walbran, P. D., et al. "Evidence from Sediments of Long-Term *Acanthaster planci* Predation on Corals of the Great Barrier Reef." *Science* 245 (1989): 847–50.

13   NUTRIA *Myocastor coypus*

Brown, L. N. "Ecological Relationships and Breeding Biology of the Nutria (*Myocaster coypus*) in the Tampa, Florida, Area." *Journal of Mammalogy* 56 (1975): 928–30.

14   HYDROZOAN *Turritopsis dohrnii*

Miglietta, M. P., and H. A. Lessios. "A Silent Invasion." *Biological Invasions* 11 (2009): 825–34.

Piraino, S., et al. "Reverse Development in Cnidaria." *Canadian Journal of Zoology* 82 (2004): 1748–54.

15   KUDZU *Pueraria montana*

Forseth, I. N., and A. F. Innis. "Kudzu (*Pueraria montana*): History, Physiology, and Ecology Combine to Make a Major Ecosystem Threat." *Critical Reviews in Plant Sciences* 23 (2004): 401–13.

Witkamp, M., M. L. Frank and J. L. Shoopman. "Accumulation and Biota in a Pioneer Ecosystem of Kudzu Vine at Copperhill, Tennessee." *Journal of Applied Ecology* 3 (1966): 383–91.

Part Four: THE WORLD OF THE FUTURE

Quammen, D. "Planet of Weeds: Tallying the Losses of Earth's Animals and Plants." *Harper's*, October 1998, 57–69.

16   ZEBRA MUSSEL *Dreissena polymorpha*

U.S. Environmental Protection Agency. "United States Great Lakes Program Report on the Great Lakes Quality Agreement." *The Great Lakes Water Quality Agreement*, July 22, 2009. www.epa.gov/grtlakes/glwqa/usreport/part5.html.

17   JELLYFISH *Phylum Cnidaria and Phylum Ctenophora*

Lynam, C. P., et al. "Jellyfish Overtake Fish in a Heavily Fished Ecosystem." *Current Biology* 16 (2006): R492–93.

Richardson, A. J., et al. "The Jellyfish Joyride: Causes, Consequences and Management Responses to a More Gelatinous Future." *Trends in Ecology and Evolution* 24 (2009): 312–22.

18   HOUSE SPARROW *Passer domesticus*

Brenckle, J. F. "Notes on the Intelligence of the House Sparrow." *Bird Banding* 7 (1936): 84.

19 WATER HYACINTH *Eichhornia crassipes*

Bicudo, Denise, et al. "Undesirable Side-Effects of Water Hyacinth Control in a Shallow Tropical Reservoir." *Freshwater Biology* 52 (2007): 1120–33.

University of Florida. *Plant Management in Florida Waters*. Center for Aquatic and Invasive Plants, University of Florida, and Florida Fish and Wildlife Conservation Commission, 2003. http://plants.ifas.ufl.edu/guide/mechcons.html.

Verma, R., S. P. Singh and K. Ganesha Raj. "Assessment of Changes in Water-Hyacinth Coverage of Water Bodies in Northern Part of Bangalore City Using Temporal Remote Sensing Data." *Current Science* 84 (2003): 795–804.

"Water Hyacinths Soak Up Pollution." *Bioscience* 26 (1976): 224.

20 NILE PERCH *Lates niloticus*

Geheb, K., et al. "Nile Perch and the Hungry of Lake Victoria: Gender, Status and Food in an East African Fishery." *Food Policy* 33 (2008): 85–89.

Yohannes, O. "Water Resources and Inter-riparian Relations in the Nile Basin: The Search for an Integrative Discourse." Albany: State University of New York Press, 2008.

CONCLUSION

"Do Biological Invasions Decrease Biodiversity? A Round Table with James H. Brown, Don. F. Sax, Daniel Simberloff and Mark Sagoff." *Conservation* 8 (2007).

Lugo, A. E "The Emerging Era of Novel Tropical Forests." *Biotropica* 41 (2009): 589–91.

Olenin, S., and E. Leppäkoski. "Non-native Animals in the Baltic Sea: Alteration of Benthic Habitats in Coastal Inlets and Lagoons." *Hydrobiologia* 393 (1999): 233–43.

PHOTO CREDITS

Page 155   © Howard Hall/Photolibrary
Page 156   © John Cancalosi/Photolibrary
Page 161   © Gerard Soury/Photolibrary
Page 165   © Spike Walker/Getty Images
Page 168   © Inga Spence/Getty Images
Page 172   © National Geographic/Getty Images
Page 177   *Top left,* © Visuals Unlimited/Corbis
           Top right, © AFP/Getty Images
           Middle left, © Peter Chadwick/Getty Images
           Middle right, © Brian S. Turner; Frank Lane Picture Agency/CORBIS
           Bottom, © AFP/Getty Images
Page 186   © Visuals Unlimited/Corbis
Page 195   © PETER YATES/SCIENCE PHOTO LIBRARY
Page 203   © Jeff Rotman/Getty Images
Page 206   © AFP/Getty Images
Page 208   © David Doubilet/Getty Images
Page 212   © Brian S. Turner; Frank Lane Picture Agency/CORBIS
Page 226   © Peter Chadwick/Getty Images
Page 229   © RUNGROJ YONGRIT/epa/Corbis
Page 235   © Reuters/CORBIS
Page 236   © Wu Dongjun/epa/Corbis
Page 240   © AFP/Getty Images
Page 245   © STEPHEN MORRISON/epa/Corbis

INDEX

squirrels and, 84, 85
Black Sea, 194, 205, 206–7
bluebirds, 217
bluegills, 34–35
Boiga irregularis. See brown tree snake
Borneo, 115
botulism, 199
box jellyfish, 202, 204, 207
Brazil, 113, 114, 237–38
Brenckle, Jacob, 219
bridled white-eyes, 36–37
British Columbia, 26, 30, 59, 74–75, 82
broom, Scotch, 75
Brown, Larry, 159–60
brown tree snake *(Boiga irregularis),* 36–46
    control measures, 45
    diet, 41–42
    impacts, 40–41
    native range, 37, 38, 42
    as predator, 40, 41
    spread, 38
bubonic plague, 22, 253
Buczkowski, Grzegorz, 144–46

C

cactus, prickly pear, 21
California
    American bullfrog, 26, 31–34
    Argentine ant, 143–44, 146–48
    Caulerpa taxifolia, 90, 95
    feral pig, 103, 105, 106, 108
    Humboldt squid, 58, 63
    jellyfish, 201
    Monterey Bay, 58, 59, 63
    native frogs, 33–34, 35, 49
    nutria, 161
    San Francisco Bay, 23, 185
    water hyacinth, 231
Canada, 103, 174, 185, 216. *See also* Great Lakes;
    specific provinces
Canada goose, 183
canals, 21, 189, 190, 191. *See also* hydrology
cannibalism, 25, 29–30
Carcinus maenas. See European green crab
Caribbean Sea, 22, 113
carp, 20, 191, 252
Caspian Sea, 194
Cassell, Scott, 62
cats, 45, 104, 160
Caulerpa taxifolia, 89–98
    control measures, 94–95
    hardiness, 89, 94

impacts, 95–98
nutrient extraction, 92, 97–98
predators, 95
reproduction, 93–94
spread, 94, 97–98
structure, 90–92
toxicity, 92, 93, 97
Central America, 52, 55, 142. *See also specific*
    countries
Cercopagis pengoi (water flea), 183–84
cheatgrass, 134
chestnut blight, 22
Chile, 59, 63, 170
China, 113, 210, 237
cholera, 128
chytrid fungus (*Batrachochytrium*
    dendrobatidis), 30–31, 47–56
    environmental change and, 51–52
    ideal conditions, 55–56
    impacts, 49
    origins, 50–52
    prevention, 56
    resistance to, 49, 54
    spread, 54–55
cichlids, 242–43, 245, 247–49
ciliates, 209
clams, 183
climate change, 51. *See also* environmental
    change
Clostridium difficile, 123–29
    antibiotic resistance, 128, 129
cnidariums (true jellyfish), 202, 204. *See also*
    jellyfish
colitis, 128–29
Cook, James, 101
Cook Islands, 102
copepods, 209
coral reefs, 151–52, 153–55
cordgrass, 161
Costa Rica, 48, 50–51, 52, 54
crabs. *See also* European green crab
    Asian shore, 67–69
    blue, 67–68
    Chinese mitten, 183, 184
    mud, 184
crayfish, 22, 184
crown-of-thorns starfish *(Acanthaster planci),*
    149–55
    diet, 151–52
    habitat and, 154–55
    impacts, 152–53
    reproduction, 154

toxicity, 151
ctenophores (comb jellies), 202, 204, 205
cuscus, 18
cyanobacteria (blue-green algae), 199, 209, 210, 238, 239, 247

D

dams, 192, 231. *See also* hydrology
Darwin's Nightmare (dir. Sauper), 245–46
dead zones, 199, 201, 205, 238
Dearborn, Ned, 216
deer, 43
deforestation, 43, 175, 180, 229, 244. *See also* forests; trees
Detroit River, 196, 197
diatoms, 209, 210, 247. *See also* plankton
dingoes, 18
diporeia, 196–97
disease, 21
    invasive species and, 22, 41, 85, 107–8, 161, 199
Dosidicus gigas. See Humboldt squid
Dreissena polymorpha. See zebra mussel
Dutch elm disease, 22

E

E. coli, 108
eagle, golden, 106
East Africa, 242, 243
eastern gray squirrel (*Sciurus carolinensis*), 79–88
    control measures, 84–85
    diet, 83, 87
    distribution, 84
    as food source, 88
    habitat and, 80, 84, 87
    impacts, 84–85
    introductions, 81
    and native species, 85, 87
    reproduction, 83
ecosystems
    forest, 180–81
    human bodies as, 123–25, 129
    invasive species and, 147–48, 250–51
    ocean, 163, 202, 204–5, 207–9
    renewal, 16, 197–98
    stressed, 43–45
    urban, 214, 222–24
egret, cattle, 252
Egypt, 231
Eichhornia crassipes. See water hyacinth
elk, Roosevelt, 75

El Niño, 63, 153–54, 208
Elton, Charles, 72, 73
England. *See* United Kingdom
environmental change. *See also* habitat disturbance
    automobiles and, 221, 223
    and bacteria, 123
    and chytrid fungus, 51–52
    natural, 16–17
    and oceans, 204–5, 209–10
    and species declines, 51, 222–23
    and species invasions, 10–11
estivation, 116
Euglandina rosea (rosy wolfsnail), 120–21
Europe, 20, 21. *See also specific countries*
    American bullfrog, 30
    Argentine ant, 141, 142, 147
    colonists from, 18–19, 22, 101
    eastern gray squirrel, 81–82
    feral pig, 102, 104, 106
    house sparrow, 220
    zebra mussel, 189, 198
European green crab (*Carcinus maenas*), 65–69
    feeding habits, 67, 68–69
eutrophication, 182, 206–7, 247
    in oceans, 205, 210
    zebra mussel and, 190, 191, 198
extinctions
    of amphibians, 49–51
    of bats, 37, 41, 44
    of birds, 37, 40, 44, 45, 185
    causes, 49, 120–21, 179
    of fish, 245
    of lizards, 37, 41, 44
    of plants, 185
    of snails, 112, 120–21

F

feral pig. *See* pig, feral
fertilizer, 56. *See also* eutrophication
fire suppression, 72, 74
fish, 245. *See also* fishing industry; *specific fishes*
    in Baltic Sea, 183–84
    in Great Lakes, 191–92, 197–98
    introductions, 20, 34–35, 191–92
    as invasive, 20–21
    jellyfish and, 207–8
fishing industry. *See also* overfishing
    European green crab and, 69
    jellyfish and, 201, 207, 208–9
    in Lake Victoria, 243, 244, 246
    zebra mussel and, 196

floodplains, 230
Florida, 21, 22
    Caulerpa, 93
    feral pigs, 101, 105, 108
    fire ants, 75
    giant African land snail, 114–15, 118
    nutria, 159–60
    water hyacinth, 238–39
food contamination, 108
food webs/chains
    in Baltic Sea, 184
    feral pig and, 106
    in freshwater bodies, 196–97, 228–29, 247
    in oceans, 209–10
forests. *See also* deforestation; trees
    feral pig and, 105–6
    kudzu and, 171, 175
    regeneration, 181
    and squirrels, 80, 82–83
    tropical, 180–81
foxes, 19–20, 76, 106
France, 30, 95, 98
frogs, 16, 34–35. *See also* American bullfrog;
      amphibians; chytrid fungus
    African clawed, 21–22, 54
    bell, 49
    declines, 30–31, 47–51, 55
    as food source, 26–28
    harlequin, 48
    mountain yellow-legged, 49
    red-legged, 33, 34, 35
fungi, 22. *See also* chytrid fungus

G

Galapagos Islands, 19, 106, 108, 178
gardens, 18–19, 223
gastrointestinal tract (human), 123, 124, 125
geckos, 135–36
Germany, 89, 185, 223
giant African land snail (*Achatina fulica*),
      111–22
    characteristics, 111–12, 115–18
    control measures, 114–15, 119–21
    diet, 112, 117–18
    as disease carrier, 118–19, 121–22
    habitat and, 114, 117
    impacts, 118, 121–22
    in native range, 112, 117–18
    reproduction, 115–16
    as resource, 114
    spread, 112–14
goats, 19, 178

gobies, 21, 184
grasses, 74–75, 132, 133–34, 138. *See also* sea
      grasses
grasslands, 105
Great Barrier Reef, 149, 153
Great Black Swamp, 191
Great Lakes, 190–93. *See also* Lake Erie; Lake
      Ontario
    mussels in, 189–90
    nonnative fish in, 21, 184, 191–92
Guam, 36–46
    invasive species, 43, 44, 46
    species declines, 37, 40–41, 43, 44, 45
Gulf of Mexico, 205, 206, 207

H

habitat. *See also* habitat disturbance
    adaptation to, 135, 137–38
    invasability of, 72
    loss of, 28, 51, 55–56, 104, 120
    new, 11
    for predators, 162
    restoration of, 35, 198
    unoccupied, 184
    urban, 84, 185
    wetland, 159
habitat disturbance, 11. *See also* agriculture;
      deforestation; hydrology
    and American bullfrog, 31–35
    and Argentine ant, 143–44
    and biodiversity, 253
    and *Caulerpa taxifolia,* 97–98
    and cichlids, 247–49
    and crown-of-thorns starfish, 154–55
    and geckos, 136
    and giant African land snail, 117
    in human body, 125, 126–28
    by humans, 10, 12, 17–23, 182, 187, 192–93
    and invasion, 72–77, 133, 135–36, 148
    by invasive species, 253
    and jellyfish, 205
    and kudzu, 174–75
    natural, 76, 134–35, 143, 182
    and plants, 74, 106, 133–34
    and water hyacinth, 228–29, 231
    and zebra mussel, 190–93
Hawaii, 136
    feral pig, 101, 104, 109
    giant African land snail, 113, 119–20
heartworm disease, 22
Hemidactylus frenatus (common house gecko),
      135–36

herbicides, 238–39
herring, 184, 191, 192, 197, 207
Hogzilla, 105
honeysuckle, Japanese, 252–53
hospitals, 129
house sparrow *(Passer domesticus)*, 213–24
  adaptability, 218–19, 220
  aggressiveness, 217
  as communal, 217–18
  declines, 213, 220–24
  diet, 214–15, 217, 219, 222
  and humans, 213–15, 218–19, 223–24
  impacts, 215–16
  nesting habits, 219, 224
  reproduction, 218, 222
  sedentary nature, 219–20, 221
  spread, 214–15, 216
  in urban environments, 214, 217, 218, 221, 222–23
Hudson River, 197
humans, 43. *See also* habitat disturbance
  Caulerpa taxifolia and, 93
  gastrointestinal tract, 123, 124, 125
  as invasion cause, 10, 12, 17–23, 18, 65–67, 179
  as invasive, 17–18
  social issues, 223–24, 244, 245–46, 253
Humboldt squid *(Dosidicus gigas)*, 57–64
  adaptability, 59
  diet, 60
  impacts, 62–63
  as predator, 60–62
  reproduction, 59–60
  spread, 57–59, 63–64
hydrozoans, 163, 164, 166–67. *See also* *Turritopsis dohrnii*
hydrilla, 228, 238–39
hydroids, 184
hydrology, 33–34, 229, 230, 231

I

India, 18–19, 120
  giant African land snail, 113, 114, 119, 120
  house sparrow, 219, 220
  water hyacinth, 231, 237
Indian Ocean, 112, 113
Indonesia, 18, 106, 112
insecticides, 44
insects, 20, 41, 134. *See also specific insects*
  declines, 222–24
  introductions, 22, 232
invasability, 72, 73

invasions, 73–74, 137, 178–79. *See also* invasability; invasive species
  enemy-release hypothesis, 132
  habitat disturbance and, 72–77, 133, 135–36, 148
  as natural, 251–52, 254
  in undisturbed areas, 73, 76
invasive species, 10, 12. *See also specific species*
  adaptability, 135–36
  and biodiversity, 95–96, 144, 179, 181, 184–85
  characteristics, 11, 132–38
  control attempts, 23, 73–74, 77, 119–21, 178, 251–52
  fitness, 132–33
  as generalists, 134–35, 194
  introductions (deliberate), 18–19, 54, 134, 137, 178
  media and, 252–53
  negative impacts, 12, 184–85, 253
  positive impacts, 12–13, 179–80, 183–84, 185, 250, 251, 252–53
  reproductive capacity, 133, 134
  spread, 10–11, 22, 23, 54, 114, 136–37
Iran, 207
Ireland, 81
islands, 106, 108, 178, 216. *See also* Pacific islands; *specific islands*
Italy, 52, 81–82

J

Japan, 113, 147, 173, 201, 204
jellyfish, 201–11
  characteristics, 202, 205
  control measures, 210–11
  diet, 202–4
  habitat and, 205
  impacts, 201, 207
  as predators, 202–4, 207–8
  as resource, 210
  spread, 201–2, 204–7, 208–9

K

Kansas, 219
Kenya, 244
King, Joshua, 75
Krakatoa, 16
kudzu *(Pueraria montana)*, 169–76
  characteristics, 171–72
  growth, 172–73
  habitat and, 174–75
  hardiness, 170

reservoirs, 237–38
rhizoids, 92
rinderpest, 22
Robison, Bruce, 63–64
Rota (island), 45
Russia, 102

S

Sagoff, Mark, 252
St. John's wort, 138
St. Lawrence River, 21, 189, 196
salamanders, 48
salmon, 191, 192, 198, 201
salvinia, giant, 228
sardines, 204
sauger, 191
Sauper, Hubert, 245–46
Savidge, Julie, 36–37, 40
Schieffelin, Eugene, 19
Sciurus carolinensis. See eastern gray squirrel
Scotland. *See* United Kingdom
sculpins, 197–98
seabirds, 40
sea grasses, 96–97, 98
sea lamprey, 191
sea nettles, 205
sea slugs, 95
sea turtle, giant, 204
sea urchins, 93
seaweed, 20, 22, 183. *See also Caulerpa taxifolia*
    Caulerpa species, 90–93, 94, 95, 97
shipping, 65–67, 190, 207. *See also* ballast
    dumping; trade
shrew, musk, 44
shrimp, 207
Simberloff, Daniel, 120
siris, white, 180
skunk, spotted, 106
slugs, 112. *See also* sea slugs; snails
smallpox, 22
smelt, 192, 197
snails, 112, 116, 117, 120–22. *See also* giant
    African land snail
    giant triton, 154
    New Zealand mud, 183, 184
    rosy wolfsnail, 120–21
    tree, 116
snakes, 22. *See also* brown tree snake
Society Islands, 121
soil erosion, 21, 173
songbirds, 84, 85
South Africa, 82, 216

South America, 17, 52–54, 113, 142, 216, 230.
    See also specific countries
South Carolina, 217
Southeast Asia, 112–13, 114, 237
South Pacific. *See* Pacific islands
sparrows, 214. *See also* house sparrow
species. *See also* invasive species
    introduced, 18, 21–22, 44, 81
    natural dispersal of, 16–17
    recovery of, 181
    specialized, 135, 192
spiders, 37, 41
squirrels, 16, 79–80, 87, 88. *See also* eastern gray
    squirrel
    and birds, 84, 85
    Delmarva fox, 16
    European red, 79–80, 85–87
starling, European, 19
stolon, 90–92
streptomycin, 128
sturgeon, lake, 191
Suarez, Andrew, 144
Suez Canal, 21
Sus scrofa. See pig, feral
Sweden, 102, 201

T

Tanzania, 55–56
Tapinoma sessile (odorous house ant), 145–46
telecommunication towers, 222
Tennessee, 106
Texas, 109, 234
tilapia, 243
toads. *See also* amphibians
    arroyo, 47
    cane, 26, 139
    golden, 50–51
    Kihansi spray, 55–56
trade, 20–21, 54, 67, 143, 144, 205–6. *See also*
    shipping
trees, 84–85, 171. *See also* forests
    Garry oak, 74–75
    on Guam, 37, 41
    paperbark, 21
    tropical, 180–81
trichomoniasis, 221
trout, 134, 154, 191, 197
Tschinkel, Walter, 75
tubeworms, 93
tuna, 63–64
Turkington, Roy, 74
Turritopsis dohrnii, 163–67. *See also* hydrozoans